Diffuse Matter in Galaxies

Cargèse 1982

NATO ASI Series

Advanced Science Institutes Series

A series presenting the results of activities sponsored by the NATO Science Committee, which aims at the dissemation of advanced scientific and technological knowledge, with a view to strengthening links between scientific communities.

The series is published by an international board of publishers in conjunction with the NATO Scientific Affairs Division

A	Life Sciences	Plenum Publishing Corporation
B	Physics	London and New York
C	Mathematical and Physical Sciences	D. Reidel Publishing Company Dordrecht, Boston and Lancaster
D	Behavioural and Social Sciences	Martinus Nijhoff Publishers
E	Engineering and Materials Sciences	The Hague, Boston and Lancaster
F	Computer and Systems Sciences	Springer Verlag
G	Ecological Sciences	Heidelberg

Series C: Mathematical and Physical Sciences No. 110

Diffuse Matter in Galaxies

Cargèse 1982

edited by

J. Audouze

Institut d'Astrophysique, Paris, France

J. Lequeux

Observatoire de Meudon, Meudon, France

M. Lévy

Université Pierre et Marie Curie, Paris, France

and

A. Vidal-Madjar

Laboratoire de Physique Stellaire et Planétaire,
Verrières-le-Buisson, France

D. Reidel Publishing Company

Dordrecht / Boston / Lancaster

Published in cooperation with NATO Scientific Affairs Division

Proceedings of the NATO Advanced Study Institute on
Diffuse Matter in Galaxies *Cargèse 1982*
Cargèse, France
September 1-16, 1982

Library of Congress Cataloging in Publication Data

Main entry under title:

Diffuse matter in galaxies, Cargèse 1982.

(NATO ASI series. Series C, Mathematical and physical sciences; no. 110)
"Published in cooperation with NATO Scientific Affairs Division."
"Proceedings of the NATO Advanced Study Institute on Diffuse Matter in
Galaxies, Cargèse, France, September 1—16, 1982" — Verso t.p.
Includes index.
1. Interstellar matter—Congresses. 2. Galaxies—Congresses. 3. Astrophysics
—Congresses. I. Audouze, Jean. II. North Atlantic Treaty Organization. Scientific
Affairs Division. III. NATO Advanced Study Institute on Diffuse Matter in Galaxies
(1982 : Cargèse, Corsica). IV. Series.
QB790.D54 1983   523.1'12   83—10844
ISBN-13:978-94-009-7185-1   e-ISBN-13:978-94-009-7183-7
DOI: 10.1007/978-94-009-7183-7

Published by D. Reidel Publishing Company
P.O. Box 17, 3300 AA Dordrecht, Holland

Sold and distributed in the U.S.A. and Canada
by Kluwer Academic Publishers,
190 Old Derby Street, Hingham, MA 02043, U.S.A.

In all other countries, sold and distributed
by Kluwer Academic Publishers Group,
P.O. Box 322, 3300 AH Dordrecht, Holland

D. Reidel Publishing Company is a member of the Kluwer Academic Publishers Group

TABLE OF CONTENTS

PREFACE

The study of the interstellar medium is certainly one of the most
recently developed astrophysical topics. Indeed, to understand
the physical status of this medium, one should observe it at
every wavelength, especially in radio and in the ranges which are
only accessible from space such as Infra Red, X and gammas.
Thanks to many new observations we now know that interstellar gas
is quite inhomogeneous, since it is made from relatively cold and
dense phases (interstellar clouds), surrounded by a relatively
hot and tenuous medium. The summer school which took place from
September 1 to 16 at the Institut d'Etudes Scientifiques of
Cargèse, and from which this book originates, was devoted to the
analysis of this second diffuse, hot and tenuous component.

The eleven sets of lectures which were presented at this school
constitute the chapters of this book. They review different
aspects (observations, determination of the physical properties
and evolution) of this diffuse medium.

In the first chapter, J. Lequeux (Marseille Observatory)
introduces the study of the diffuse interstellar medium which
plays a major role in galactic evolution. His contribution
provides some historical references on this subject, such as a
copious bibliography on the analysis and observations of the
interstellar medium.

The second contribution, supplied by M. Jura (UCLA), is a review
of the spectroscopic analysis of the absorption lines of the
interstellar medium. In this chapter one is made aware of the
considerable impact of Copernicus in this domain, as well as of
the importance and relevance of future use of the High Resolution
Spectrograph (HRS) of the Space Telescope to perform the analyses
of these absorption lines.

T. de Jong (Amsterdam Observatory) then provides an account of
the interstellar H_2 molecules emphasizing in particular its
formation and destruction processes; he also makes some remarks
on the presence of other interstellar molecules and grains.

J. Audouze et al. (eds.), Diffuse Matter in Galaxies, vii–ix.
Copyright © 1983 by D. Reidel Publishing Company.

The contribution of H.J.G.L.M. Lamers (Astronomy Institute of Utrecht) is split into two separate chapters. The first one deals with the fundamental problem of the stellar winds and their interaction with the interstellar medium. This problem is currently the subject of a great number of studies, which are extremely well summarized in this chapter. The second chapter examines the influence of supernovae explosions and remnants on the interstellar medium, and how we measure the importance of the phenomena according to the energy produced by these explosions and their objects.

Following this description of the structure and energy sources of the interstellar medium, the next two contributions concern nucleosynthesis and the compositional evolution of the inter- stellar medium, and its cosmological relevance. A. Vidal-Madjar (LPSP, Verrières-le-Buisson) devotes his chapter to pointing out the considerable importance of the interstellar abundances of D and He, which are cosmological "monitors" as critical as the 3K radiation and the expansion of the Universe. J. Audouze (Institut d'Astrophysique, Paris) reviews the principal formation mechanisms of the chemical elements, as well as some of those which govern its chemical evolution.

Following this is the very important chapter written by B. Lazareff (University of Berkeley and University of Grenoble) on the dynamics and energetics of the interstellar medium. This valuable contribution provides the reader with the necessary physical bases concerning radiation energy losses, the ionization processes and the thermal conduction of the interstellar gas, as well as the processes governing the dynamic evolution of the interstellar medium. Then, M. Perinotto (Arcetri Observatory) describes the basic theoretical and observational points leading to an understanding of the optical and infrared emission of the interstellar gas.

In chapter 9, D.R. Flower (Durham University) provides the elements of atomic physics important to the understanding of the diffuse interstellar medium. He describes in particular the transfer, the dielectronic recombination, the photoionization and radiative recombination reactions. The two last chapters were written by M. Grewing (Tubingen University) and A. Boksenberg (Royal Greenwich Observatory): chapter 10 describes the overall electromagnetic radiation emitted or absorbed by the interstellar medium, while chapter 11 provides a quick summary of present knowledge on the galactic halos and the intergalactic medium.

We hope that our readers, through this book, wll share some of the pleasure we had (lecturers and participants alike) in experiencing the scientific atmosphere of this school. For this we thank the authors of these chapters, especially those who were

kind enough to send us their contribution in time! We acknowledge
all the time and effort they put into their presentations and
their writing. We also thank the participants, who made this
school a very lively and productive one by their penetrating
questions and comments.

This Cargèse summer school has benefited from a large number of
very favourable circumstances. First, the financial help from
NATO is very gratefully acknowledged. We want to thank very
warmly Jean-Louis Basdevant who advised and encouraged us during
the organization of this session. We express our gratitude to
Professor Maurice Lévy who let us use the marvelous Cargèse
Institute. We also thank the CNES for financial contribution.

The success of this session is not only due to the scientists who
took part in it, but certainly also to those (Marie-France
Hanseler, Mary McClean and Gérard Denis Bernia) who made the
atmosphere of it extremely cheerful and pleasant.

We hope that all these individuals will find here the expression
of our gratitude.

    Jean Audouze   James Lequeux   Alfred Vidal-Madjar

P.S. We regret the absence of contributions by MM. A. Boksenberg,
     M. Grewing and T. De Jong. Unfortunately the lecture notes
     of these individuals were never received.

PREFACE

L'étude du milieu interstellaire est certainement l'un des sujets en astrophysique dont le développement est très récent. En effet, pour parvenir à comprendre sa physique, il faut l'observer dans tous les domaines de longueur d'onde en particulier en radioastronomie et dans ceux qui ne sont accessibles que de l'espace comme l'Infra Rouge, l'UV, les X et les gamma. Grâce à ces observations, on sait que c'est un milieu très hétérogène, constitué de phases relativement froides et denses (les nuages), entourées d'un milieu relativement chaud et ténu. L'école d'été qui s'est tenue à l'Institut d'Etudes Scientifiques de Cargèse du 1er au 16 septembre et dont ce livre rend compte a porté sur l'étude de cette deuxième composante diffuse, chaude et de faible densité. Les onze cours présentés à cette école, qui constituent les chapitres de ce livre, passent en revue différents aspects portant sur l'observation, la détermination des propriétés physiques et l'évolution de ce milieu diffus.

Dans le premier chapitre, J. Lequeux (Observatoire de Marseille) présente une introduction à l'étude de ce milieu interstellaire diffus qui joue un rôle majeur dans l'évolution des galaxies. Sa contribution fournit des repères historiques concernant ce sujet ainsi qu'une abondante bibliographie sur la façon d'analyser ce milieu interstellaire ainsi que sur les principaux résultats d'observation.

La deuxième contribution constitue une revue des analyses spectroscopiques des raies d'absorption du milieu interstellaire ; elle est due à M. Jura (UCLA). On remarque l'apport considérable de "Copernicus" dans ce domaine, ainsi que l'intérêt qu'il y aura d'utiliser le spectrographe à haute résolution (HRS) du Télescope Spatial pour ces analyses.

T. de Jong (Observatoire d'Amsterdam) expose ensuite tout ce qui concerne la molécule H_2 interstellaire en particulier les processus qui la forment et la détruisent sans oublier de mentionner très rapidement quelques implications de la présence des autres molécules et des grains interstellaires. La contribution de

J. Audouze et al. (eds.), Diffuse Matter in Galaxies, xi–xiii.
Copyright © 1983 by D. Reidel Publishing Company.

H.J.G.L.M. Lamers (Institut d'Astronomie d'Utrecht) se subdivise
en deux parties. La première évoque la question fondamentale des
vents stellaires et de leur interaction avec le milieu interstellaire : il est indéniable que ce problème est maintenant l'objet
de nombreuses études qui sont excellement résumées dans cette
contribution. Le second traite de l'influence des explosions de
supernovae et des restes de supernovae sur le milieu interstellaire. Là aussi on mesure l'importance de ce phénomène en raison
de la contribution énergétique de ces explosions et ces objets.

Après cette description de la structure et des sources d'énergie
du milieu interstellaire, les deux contributions suivantes dues
respectivement à A. Vidal-Madjar et J. Audouze concernent la nu-
cléosynthèse et l'évolution de la composition du milieu interstellaire. La première s'attache à relever l'importance considérable des abondances interstellaires de deutérium et d'hélium qui
sont des "moniteurs" cosmologiques des galaxies. La seconde passe
en revue les principaux mécanismes de formation des éléments chi-
miques constituant le milieu interstellaire ainsi que ceux qui
influent sur son évolution chimique.

On trouve ensuite dans le livre le "plat de résistance" constitué
par le très important chapitre rédigé par B. Lazareff (Berkeley
et Université de Grenoble) sur la dynamique et l'énergétique du
milieu interstellaire. Cette contribution donne au lecteur les
bases physiques nécessaires concernant les pertes d'énergie par
rayonnement, les processus d'ionisation et la conduction thermi-
que de ce gaz et tous les processus d'évolution dynamique du milieu interstellaire. Puis M. Perinotto (Observatoire d'Arcetri)
décrit les éléments théoriques et observationnels à partir des-
quels on peut comprendre l'émission du gaz interstellaire en op-
tique et dans l'Infra Rouge.

Dans le chapitre 9, D.R. Flower (Université de Durham) fournit
les rudiments de physique atomique applicables à la physique du
milieu interstellaire diffus. On y trouve, en particulier, une
description des réactions de transfert, de recombinaison ou di-
électronique de photoionisation et de recombinaison radiative.
Les deux derniers chapitres sont dûs à M. Grewing (Université de
Tubingen) et A. Boksenberg (Royal Greenwich Observatory). Le chapitre 10 décrit l'ensemble du rayonnement électromagnétiue émis
ou reçu par le milieu interstellaire, tandis que le chapitre 11
s'attache à présenter un résumé des connaissances actuelles sur
les halos galactiques et le milieu intergalactique.

Nous espérons que les lecteurs de ce livre auront un peu du plai-
sir qu'ont eu l'ensemble des participants (y compris les confé-
renciers) à goûter l'atmosphère scientifique de cette école. Nous
remercions pour cela les auteurs des différents chapitres et surtout ceux qui les ont remis à temps pour toute la peine qu'ils

ont pris à rédiger leur contribution ainsi que l'ensemble des
participants pour avoir animé les exposés grâce à leurs questions
et commentaires pénétrants.

Cette école d'été de Cargèse a bénéficié d'un grand nombre de
circonstances très favorables : Tout d'abord l'aide financière
qui nous fut prodiguée par l'OTAN et dont nous remercions cet
organisme. Nous tenons à remercier très vivement Jean-Louis
Basdevant pour nous avoir conseillé et encouragé dans le cours de
l'organisation de cette session scientifique. Nous voulons expri-
mer notre gratitude au Professeur Maurice Lévy qui a mis à notre
disposition ce merveilleux Institut de Cargèse. Nous remercions
également le C.N.E.S. pour son aide financière.
Le succès de cette session ne revient pas seulement aux scientifiques qui y ont pris part, mais peut-être et surtout à celles
(Marie-France Hanseler et Mary McClean) et celui (Gérard Denis
Bernia) qui ont su rendre l'atmosphère de cette session particu-
lièrement agréable et chaleureuse.

Que tous trouvent ici l'expression de notre reconnaissance.

Jean Audouze   James Lequeux   Alfred Vidal-Madjar

P.S. Nous regrettons l'absence des contributions de
     MM. A. Boksenberg, M. Grewing et T. De Jong, qui ne
     nous sont pas parvenues.

LIST OF PARTICIPANTS

Lecturers :

J. AUDOUZE, Institut d'Astrophysique, Paris, France
A. BOKSENBERG, Royal Greenwich Observatory, Hailham, U.K.
D. FLOWER, University of Durham, Durham, U.K.
M. GREWING, Astronomisches Institut, Tubingen, FRG
T. DE JONG, Astronomical Institute, Amsterdam, Netherlands
M. JURA, Astronomy Dept., U.C.L.A., Los Angeles, USA
H.J.G.L.M. LAMERS, Van Houtenweg 4, De Bilt 3732BF, Netherlands
B. LAZAREFF, Astronomy Dept., Berkeley, USA
J. LEQUEUX, Observatoire de Meudon, Meudon, France
M. PERINOTTO, Observatorio Astronomico di Arcetri, Firenze, Italy
A. VIDAL MADJAR, Physique Stellaire et Planétaire, Verrières-le-
                    Buisson, France

Participants :

L.J. ALLAMANDOLA, Lab. of Astrophysics, Leiden, Netherlands
A. ALTAMORE, Istituto Astronomica, Roma, Italy
K.R. ANANTHARAMAIAH, Raman Research Inst. Bangalore, India
M. ARNAUD, Astrophysique, C.E.N. de Saclay, Gif-sur-Yvette, France
P. BARGE, Astronomie Spatiale, Marseille, France
P. BEDIJN, Theoretische Astrophysik, Heidelberg, FRG
J. CAPLAN, Observatoire de Marseille, France
A. CASTETS, Astrophysique, CERMO, Grenoble, France
M. CERRUTI-SOLA, Osservat. Astrofisico di Arcetri, Firenze, Italy
P. COX, Observatoire de Marseille, France
Y. DANIELS, Nuffield Radio Astronomy lab., Jodrell Bank, U.K.
A.I. DIAZ, Royal Greenwich Observatory, Hailham
J.R. DUCATI, Centre de données stellaires, Strasbourg, France
R. FERLET, ESO, Garching, FRG
E.L. FITZPATRICK, Washburn Observatory, Madison, Wisconsin, USA
S.K. GHOSH, Infrared Astronomy Group, Tata Institute, Bombay,Inde
A.A. GOHARDJI, Physics Dept. University, Durham, U.K.
M. GRAFF, Oregon University, Eugene, USA
C. GRY, Physique Stellaire et Planétaire, Verrières-le-Buisson, F
D. INNES, Astrophysics, University College, London, U.K.

N.E.JAIDANE, Dép.de Physique, Univ.de Tunis-le Belvédère, Tunisia
M.F.KESSLER, Astronomy, European Space Agency, Noordwijk, Netherl.
J. KREBS, Max Planck Institut für Astrophysik, Garching, FRG
A. LEGER, Physique des Solides, ENS, Paris, France
L. LE SERGEANT D'HENDECOURT, Huygens Lab. L.A.F. Leiden, Netherl.
LU JUFU, International Center for Theoretical Phys. Trieste,Italy
G. MALINIE, Institut d'Astrophysique, Paris, France
F. MARCHESONI, Istituto di Fisica, Pisa, Italy
M. de MUISON, Astronomie Infrarouge, Observatoire de Paris, Meudon
M.J. MURRAY, Physics Dept. Queen's University, Belfast, Irland
N. PANAGIA, Radioastronomia, CNR, Bologna, Italy
R.T.H. PWA, Kapteyn Lab. Groningen, Netherlands.
T. SABOURIN, Astrophysique, Bordeaux, France
L. SANZ FERNANDEZ DE CORDOBA, Madrid, Spain
E. SCHULTZ-LUPERTZ, Astronomisches Inst. Tübingen, FRG
F. STRAFELLA, Dip. di Fisica, Lecce, Italy
J. SPICKER, Astronomisches Inst. Bochum, FRG
I. TARRAB, Institut d'Astrophysique, Paris, France
P. TOMASI, Radioastronomia, Bologna, Italy
M. TORISEVA, Astrophysics Lab., University, Helsinki, Finland
M. TOSI, Osservatorio Astronomico Universitario, Bologna, Italy
A. TREW, Royal Observatory, Edinburg, U.K.
J. TSEZOS, Astrophysics, National University, Athens, Greece
M.S. UNGER, Nuffield Radio Lab., Jodrell Bank, Macclesfield, U.K.
S. VAUCLAIR, Observatoire de Toulouse, France
L. WESTERN, Loomis Lab. of Physics, Urbanna, Ill, USA
M. WINTHER, Astronomisk Institut, Aarhus, Danmark
H. ZEKL, Institut für Theoretische Physik, Frankfurt, FRG

GENERAL INTRODUCTION TO THE DIFFUSE INTERSTELLAR MEDIUM ; RADIO
OBSERVATIONS

James Lequeux

Observatoire de Meudon, 92190 Meudon, France
also at Observatoire de Marseille

I . IMPORTANCE OF THE INTERSTELLAR MEDIUM ; BRIEF HISTORY OF
    INVESTIGATIONS

1 . The importance of the interstellar medium (ISM)

This importance is twofold. i) It plays a major role in the
evolution of galaxies. Galaxies contain up to 40 per cent of
their mass as interstellar gas and dust. Stars are born from this
interstellar medium and return continuously to it matter enriched in heavy elements by nucleosynthesis. Moreover exchanges
of interstellar gas between different parts of a galaxy and exchanges between a galaxy, its neighbours during encounters and
with the intergalactic medium are likely to modify strongly this
simple picture of galactic evolution. ii) The ISM is a very
interesting laboratory of physics and physico-chemistry showing a
vast range of physical parameters. A large variety of interesting
phenomena can be studied in the ISM, often in conditions far from
thermodynamic equilibrium : formation and destruction of molecules, ionization and recombination,excitation of atomic and
molecular levels and line formation, condensation and destruction
of grains, thermal exchanges, hydrodynamic phenomena (including
shocks), etc. The domain of interstellar studies is thus a carre-
four of astrophysics and physics: this explains why it is presently one of the most active parts of astronomy.

Roughly speaking, the ISM can be separated into two cate-
gories of objects with very different physics : the diffuse medium, which is the subject of the present school, and the denser
molecular clouds which will not be studied here, but at the 1983
session of the Les Houches School of Theoretical Physics.

1

J. Audouze et al. (eds.), Diffuse Matter in Galaxies, 1–18.
Copyright © 1983 by D. Reidel Publishing Company.

2. A brief history of interstellar observations

In this paragraph and in the following one, I give very few references ; most can easily be found in textbooks and review
papers, in particular in Middlehurst and Aller (1968) for earlier
work.

The first decisive step in our knowledge of the ISM is the
discovery by William Huggins (1864) of emission lines in the
Orion Nebula and his acknowledgement that some lines (hydrogen
lines!) were at the same wavelength as lines emitted by excited
gas in the laboratory. This established that at least some "nebulae" were made of (hot) gas ; others like the Andromeda Nebula
did not show emission lines (at least with the instruments of the
time) and were rightly considered as made of stars. In 1904,
Hartmann observed that the CaII resonance lines (H and K) in the
spectrum of the spectroscopic binary δ Ori had constant wavelengths, contrary to the stellar lines, and suggested that they
were formed in some steady gas along the line of sight ; this
idea was developed by Slipher (1909) then by Miss Heger (1919)
who discovered the similar NaI D lines. The idea of an ISM producing those lines evolved apparently very gradually, and culmin-
ated with the high-resolution work made in 1930-1950 by Wilson,
Merrill, Adams, Dunham and others, who showed the complexity of
these interstellar lines, revealing the clumpy nature of the ISM,
and discovered IS lines from other atoms and ions (TiII, CaI, KI,
FeI). Some other lines were shown by Swings and Rosenfeld, McKel-
lar, Douglas and Herzberg in 1937-1942 to come from diatomic
molecules (CN, CH, CH+). Other absorption features, broad and
shallow ("bands"), obviously also interstellar in origin, were
discovered by Merrill already in 1934 but are still unidentified.

Meanwhile, further studies of emission nebulae revealed
lines unknown in the laboratory, which were only identified correctly by Bowen in 1927 as forbidden lines of common ions.

The ISM affects the light of stars in different ways (extinction, scattering, polarization). The existence of dark
"clouds" was well documented thanks to the observations of Bar-
nard, Max Wolf and Curtis early in this century, but their interstellar nature was realised only slowly and Trümpler was the
first to show in 1930 that the light of distant stars is most
generally affected by extinction due to IS dust. Slipher (1912)
discovered that a nebula near the star Merope had the same spectrum as Merope itself and Hertzsprung correctly interpreted this
by diffusion by dust particles ; the link between these "reflex-
ion nebulae" and the dark clouds was later realised rather slowly.

In 1951, the IS line of atomic hydrogen at 21-cm wavelength
was discovered by Ewen and Purcell, following the prediction of

van de Hulst, and opened a new dimension in IS studies by offer-
ing for the first time the way of observing the main constituent
of the ISM and of determining its temperature (see later).
Zeeman 21-cm observations allowed the first measurement of the
interstellar magnetic field (Verschuur, 1968), the existence of
which was infered from other observations (interstellar polarization resulting from grain alignment by the magnetic field, galactic synchrotron radiation). Other important discoveries due to
radioastronomy in the 1950's were those of the thermal free-free
emission of HII regions, of the non-thermal (synchrotron) emission of supernova remnants (Shklovsky, Alfven) and of the whole
Galaxy and external galaxies ; in the early 1960's were discovered recombination lines emitted by HII regions and even HI regions. We will come back to these points later. Radio observations of IS molecules (OH in 1963 : Weinreb, Barrett, Meeks and
Henry, OH masers in 1965, NH_3, H_2O masers, and finally CO and
many molecules discovered by millimeter-wave emission from 1969
on by Penzias, Wilson and a bunch of others) opened another
window in IS studies, emphasizing the enormous importance of
molecular ($\equiv$ dark) clouds which make roughly half of the mass of
the ISM in our Galaxy, and where stars are born.

    Space research provided an enormous observational input to
our knowledge of the ISM, which is indeed observable in all
ranges of wavelengths. Many of the most interesting results come
from UV observations : discovery of IS molecular hydrogen (Car-
ruthers, 1969), discovery and abundances of a very large number
of IS atoms and ions through far-UV absorption lines, including
very highly ionized species like CIV, NV, OVI (Copernicus satellite, after 1973), and discovery of the hot galactic halo
(Savage and de Boer with the International Ultraviolet Explorer,
1981). IR observations are not less exciting, since they reveal
the dust thermal emission (half of the power emitted by our
Galaxy is far-IR emission by dust heated by stellar light) and a
bunch of very important atomic, ionic and molecular lines. X-rays
are essential to understand supernova remnants and the hot
regions of the ISM in general, while gamma-ray observations are
quite important to map indirectly the ISM in remote regions of
our Galaxy. Needless to say that many of the observations and
discoveries made in our Galaxy have their equivalent in other
galaxies, except when sensitivity does not allow them (for
example we still have no gamma-ray or high-dispersion UV spectroscopy in external galaxies).

3. Evolution of the concept of interstellar medium

Until the 1960's, our ideas about the ISM were still extremely
primitive, mainly due to the lack of observations. The only field
which was really well developed was the theory of excitation of
HII regions by the ionizing flux of a central star (the idealized

"Strömgren sphere") : this is due to Bowen (1923), Zanstra (1925)
and Strömgren (1939) while Menzel (1937-1945) gave the theory of
emission of recombination lines (our complete understanding of
the forbidden lines came later and is mainly the work of Seaton
and collaborators in London). The status of our knowledge in this
period is extensively described in Vol. VII of the compendium
"Stars and Stellar Systems" (Middlehurst and Aller, 1968), whose
material mostly dates from before 1960-63. The average density of
the ISM in the Galaxy was known to be $\approx$ 1 H-atom cm^{-3} from 21-cm
observations ; the abundance of heavy elements in the ISM is
similar to that in the Sun or most stars (from analysis of the
composition of the Orion Nebula), but some elements like Ca and
Ti are clearly depleted in the clouds as shown by IS line observations. Most of the ISM was supposed to be concentrated in clouds
with radius $\approx$ 7 pc, density $\approx$ 20 atom cm^{-3} and mass $\approx 10^3$ M$_{\odot}$. The
concept of a "standard cloud" resulting from statistics on 21-cm
line, interstellar absorption lines and interstellar extinction
was generally accepted, however some authors advocated for clouds
of much larger masses and sizes (essentially the dark clouds).
The HI (neutral) clouds were realized to be opaque to Lyman continuum photons but more or less transparent to photons with $\lambda >$
912 A, resulting in the ionization of all elements with an
ionization potential < 13.6 eV (i.e. C, metals, etc.). The observed cloud temperatures of some 50 K or more (from a discussion
of the 21-cm emission-absorption observations) did not appear to
create much theoretical problem since inelastic cloud collisions
and cosmic rays were supposed to do the heating ; the temperature
of $\approx 10^4$ K observed for HII regions was already well understood.
The dynamics of the ISM was still in an extremely rudimentary
state, due both to the lack of key observations and to the un-
satisfactory status of the necessary physics (see, e.g. Spitzer,
1968). Although some scientists like Field, Kahn, Kaplan, Schatz-
man, Spitzer and others had penetrating insights, only some
simple problems could be considered as having received the beginning of a solution. It was realized however that IS clouds have
to be maintained by some sort of external pressure. In the
galactic disk, a general intercloud gas ionized by hot stars was
usually believed to do the job. For the IS clouds discovered high
above the galactic plane by Münch (optical absorption lines) and
by Oort and collaborators ("high velocity clouds" seen in 21 cm),
Spitzer (1956) proposed that they are maintained by a corona of
very hot ionized gas (T $\approx 10^6$ K, $n_e \approx 10^{-3}$ at cm^{-3}) : this idea
was proven to be true only very recently by the direct detection
of this corona (Savage and de Boer, 1981).

A significant progress in the concept of the ISM is the two-
component model of Field, Goldsmith and Habing (1969) (although
this model had to be deeply revised subsequently). As indicated
above, the idea of a cloud population in equilibrium with an
intercloud medium was not new but the authors identified the

intercloud medium with a hot ($\gtrsim 10^3$ K) <u>neutral</u> HI gas evidenced
by 21-cm observations, and, making a stability analysis, showed
that the existence of the two stable phases in pressure equilibrium (cold clouds and hot neutral intercloud gas) was natural if
the ISM was heated by low-energy cosmic-rays. They also explicit-
ly predicted the possibility of a third stable phase above 10^6 K,
with bremsstrahlung as the chief (slow) cooling process but this
phase was not yet observed. In this model, HII regions play only
a marginal role.

Several new elements have yielded major (almost complete)
changes in this picture. First there is no such thing as a
standard IS cloud : they come in a variety of sizes and masses;
however this has little impact on the model since it turns out
that the measured properties of actual HI clouds imply that they
may well be in equilibrium with an intercloud medium of external
pressure n k T $\approx$ 2000 k cm^{-3} K whatever their size. Second,
observations show that the ionization rate inside clouds is not
compatible with cosmic-ray heating, which would require an
ionization rate $\zeta \approx 10^{-15}$ s^{-1} per H-atom instead of the observed
$\approx 10^{-17}$ (Watson, 1975, see p. 276). One must then look for other
heating mechanisms : soft X-ray heating being questionable at
least for clouds, photoemission by dust grains and perhaps
formation on grains of H_2 molecules released with appreciable
kinetic energy are presently considered as the most efficient
ones in general. This is not sufficient, however, to kill the
standard two-component medium. However, some of the foundations
were shaken and the paper by Field (1975) expresses well the
uncertain state of affairs at that time.

A major blow came from the recognition that a major fraction
of the ISM should be filled with low-density hot "coronal" gas,
at least in the solar neighbourhood : this comes from the observations of ubiquitous OVI far-UV absorption lines and even more
directly from the observations of soft X-rays emitted by this
medium ; also, 21-cm pictures show that the neutral ISM is
extremely irregular and shows conspicuous bubbles, sheets, arches,
etc. It is now recognized that this is a consequence of supernova
explosions and of winds emitted by massive hot stars. Consequently the ISM is in a state of permanent buoyancy and has the
structure of a "swiss cheese" where the cavities are irregular
tunnels formed by merged supernovae (and stellar wind) cavities
(McCray and Snow, 1979 ; McKee and Hollenbach, 1980). It is no
more possible to consider stationary models for the ISM. This
picture was just beginning to emerge in 1974 and is at the basis
of the 3-component model of the ISM (McKee and Ostriker, 1977).
In this model, " supernova explosions in a cloudy ISM produce a
three-component medium in which a large fraction of the volume is
filled with hot, tenuous gas. In the disk of the Galaxy the
evolution of supernova remnants is altered by evaporation of cool

clouds embedded in the hot medium... Very small clouds will be
rapidly evaporated or swept up. The outer edge of clouds ionized
(and heated) by diffuse UV and soft X-ray backgrounds provide the
warm ($\approx 10^4$ K) ionized and neutral components." Although the
model is not static, it is still in pressure equilibrium since
the sound travel-time through all components is shorter than all
other characteristic times. Stellar winds (Castor et al., 1975 ;
Weaver et al., 1977), ignored by McKee and Ostriker, produce a
further energy input but should not change the picture drastically. In regions of small densities without many neutral clouds,
supernova remnants and stellar winds from individual stars or
associations result in conspicuous bubbles seen in the 21-cm line
and in Hα light. Expansion of HII regions in such an irregular
medium is an interesting problem presently under study.

Although generally convincing, the 3-component model should
not be taken too literally however. For example, the classical
warm intercloud medium of Field et al. could well be widespread
in regions not heated by supernova explosions. On the other hand,
repeated supernova explosions in stellar associations may
dissipate entirely the cool and warm gas and induce the formation
of very large holes such as observed in our Galaxy and in some
other galaxies, which are probably filled only with hot gas (see
e.g. Bruhweiler et al., 1980). If most of the massive stars
evolve in large groups, as indeed suggested by observation, the
swiss-cheese picture of McKee and Ostriker should be modified
appreciably, and a galaxy would rather consist of these giant
holes separated by 2- or 3-component regions. This remains to be
checked by detailed studies of external galaxies, if possible in
soft X-rays.

The situation is somewhat different in the galactic corona.
The hot gas is not confined to the disk but tends to flow away
from it ; as the gas flows outwards it cools slowly and thermal
instability forms denser, neutral clouds which are called back to
the galactic plane by gravity (Shapiro and Field, 1976) ; these
clouds might be the High Velocity Clouds observed at 21-cm. This
is the galactic fountain model.

Much interest exists at present in the time evolution of the
ISM. It is clear that IS clouds must evolve continuously : they
form from thermal instabilities and by the compression of dense
cold shells in old supernova remnants and accrete material from
these shells. On the other hand they continuously evaporate into
the hot ISM. Cloud collisions result in partial coalescence and
partial fragmentation. When the clouds are big enough, supersonic
collisions produce compressions in which atoms are converted into
molecules : molecular clouds are thus formed and are stabilized
by gravity rather than by external pressure as are the atomic
clouds ; this occurs for masses $\approx 10^3$ M$_\odot$. Further on, the

molecular clouds are disrupted by the massive stars they form and
their matter returns into the diffuse medium. Modelisation of
these complex phenomena is a formidable problem which has been
attacked only partially ; moreover, in a real spiral galaxy,
compression at the spiral arms introduces another element in the
dynamical evolution which is of obvious importance for the
formation of molecular clouds. Partial results have been obtained;
e.g. the steady-state mass spectrum of the diffuse clouds has
been studied by Chièze and Lazareff (1980) in the frame of the 3-
component model, and the formation of molecular clouds in the
spiral arms by Casoli and Combes (1982) and Boulanger (1982). The
dynamical evolution of HII regions and their effect on the ISM
are studied by Tenorio-Tagle and collaborators (see, e.g. Tenorio-Tagle et al., 1979), and others.

Most of the previous works have ignored the magnetic field.
Although this neglect is certainly not justified in general, the
bad problem is that one can obtain essentially any result one
wishes by choosing an appropriate topology for the magnetic
field. However there is one well-studied case where magnetic
field probably plays a role in the evolution of the ISM : that of
the Parker instability (Parker, 1966) which may help the galactic
fountain mechanism and also contribute with self-gravity to the
formation of large cloud assemblies in spiral arms (see e.g.
Elmegreen, 1982).

References

Balian, R., Encrenaz, P., Lequeux, J., ed., 1975. Atomic and
     Molecular Physics and the Interstellar Matter, North Holland
     Public. Co., Amsterdam
Boulanger, F., 1982, in preparation
Bruhweiler, F.C., Gull, T.R., Kafatos, M., Sofia, S., 1980,
     Astrophys. J. Letters, 238, pp. L27-30
Casoli, F., Combes, F., 1982, Astron. Astrophys., 110 pp.287-294
Castor, J., McCray, R., Weaver, R., 1975, Astrophys. J. Letters,
     200, pp. L107-110
Chièze, J.-P., Lazareff, B., 1980, Astron. Astrophys., 91, pp.290
     -301
Elmegreen, B., 1982, Astrophys. J., 253, pp.655-665
Field, G.B., 1975, in Balian et al., op.cit., pp.467-531
Field, G.B., Goldsmith, D.W., Habing, H.J., 1969, Astrophys.J.
     Letters, 155, pp. L149-154
McCray, R., Snow, T.P., 1979, Ann. Rev. Astron. Astrophys., 17,
     pp. 213-240
McKee, C.F., Hollenbach, D.J., 1980, Ann. Rev. Astron. Astrophys.
     18, pp.219-262
McKee, C.F., Ostriker, J.P., 1977, Astrophys. J., 218, pp.148-169
Middlehurst, B.M., Aller, L.H., 1968. Nebulae and Interstellar
     Matter (Stars and Stellar Systems, Vol. VII), the University

of Chicago Press, Chicago

Parker, E.N., 1966, Astrophys. J., 145, pp. 811-833

Savage, B.D., de Boer, K.S., 1981, Astrophys. J., 243, pp. 460-
    484

Shapiro, P.R., Field, G.B., 1976, Astrophys. J., 205, pp.762-765

Spitzer, L., 1956, Astrophys. J., 124, pp. 20-34

Spitzer, L., 1968 in Middlehurst and Aller, op.cit., pp. 1-63

Tenorio-Tagle, G., Yorke, H.W., Bodenheimer, P., 1979, Astron.
    Astrophys., 80, pp. 110-118

Watson, W.D., 1975, in Balina et al., op. cit., pp. 179-324

Weaver, R., McCray, R., Castor, J., Shapiro, P., Moore, R., 1977,
    Astrophys. J., 218, pp. 377-395

II. BRIEF SUMMARY OF THE METHODS OF INVESTIGATION OF THE DIFFUSE INTERSTELLAR MATTER, WITH REFERENCES

1. Selected books on the interstellar medium

- Middlehurst and Aller, 1968 : Nebulae and Interstellar Matter (see references). This book is now partly of historical interest but contains nevertheless many chapters which are still basic today.

- Verschuur, G.L., Kellermann, ed., 1974 : Galactic and Extragalactic Radio Astronomy, Springer Verlag, Berlin. This is still the best book giving the basis of radioastronomical studies of the ISM, although many items are not up to date.

- Osterbrock, D.E., 1974 : Astrophysics of Gaseous Nebulae, Freeman and Co., San Francisco. The classical book on the physics of HII regions and planetary nebulae. Could be complemented from an illustrative example by :

- Goudis, C., 1982 : The Orion Complex : a Case Study of Interstellar Matter, D. Reidel, Dordrecht. Many photographs and drawings ; useful compendium of general formulae and numerical data.

- Balian, R., Encrenaz, P., Lequeux, J., ed., 1975: Atomic and Molecular Physics and the Interstellar Matter (see refer- ences). This Les Houches summer course is another basic book ; see in particular the chapters by Watson (chemistry), Field (structure and dynamics) and Kahn (hydrodynamics).

- Spitzer, L. Jr., 1978 : Physical Processes in the Inter- stellar Medium, Wiley, New York. The most basic reading on the ISM, in particular diffuse. Most of the necessary physics is given in a very comprehensive and concise way ; a carefuly reading is really necessary to fully benefit from this book,

which replaces the earlier version Diffuse Matter in Space,
published in 1964.

- Dyson, J.E., Williams, D.A., 1980 : The Physics of the
Interstellar Medium, Manchester University Press. A good general
book of lower level, with emphasis on dynamics.

2. Methods of investigation of the diffuse ISM

I will go rapidly through the different wavelengths, emphasizing
what is their major interest for ISM.

- Gamma rays : produced by interaction of cosmic rays with
the ISM. Give total amount of ISM assuming cosmic-ray flux.
Limited sensitivity and resolution prevents their use for detailed
studies of the diffuse ISM. See e.g. Cesarsky, C.J., Paul, J.A.,
1981, Nuclear Science and Applications, 1, pp.191-225.

- X-rays : X-ray absorption at low energies $\lesssim 1$ keV gives
the total amount of ISM in front of the studied X-ray source
(e.g. Ryter, C., Cesarsky, C.J., Audouze, J., 1975, Astrophys. J.
198, 103). X-rays are emitted by bremsstrahlung by hot gases ($T \gtrsim$
10^6 K) and are fundamental tracers of the hot ISM, however
plagued by the above absorption : soft X-ray background, supernova remnants (Gorenstein, P., Tucker, W.H., 1976, Ann. Rev.
Astron. Astrophys., 14, pp. 373-416). X-ray lines produced by
highly ionized ions (OVII, OVIII, FeXVII, NeIX, etc.) are well
observed in supernova remnants and start to be observed in the
hot diffuse ISM (Schnopper et al., 1982, Astrophys. J., 253, pp.
131-135) : they are very promising monitors of temperature and
density.

- Ultraviolet : The diffuse UV background is believed to be
in part galactic light scattered by high-latitude dust and gives
information on this dust ; observation of line emission would be
very interesting for hot and warm ISM (Paresce, F., Jakobsen, P.,
1980, Nature, 288, pp. 119-126 ; Jakobsen, P., Paresce, F., 1981,
Astron. Astrophys., 96, 23, etc.). Ultraviolet extinction and
reflected UV light are fundamental for our understanding of IS
dust (Savage, B.D., Mathis, J.S., 1979, Ann. Rev. Astron. Astrophys., 17, pp. 73-111). IS absorption lines are extremely abundant and have really renewed our vision of the diffuse ISM : H
and D, H_2 and HD, depletion of elements, highly ionized ions like
CIV, NV, OVI, lines from excited levels giving electron densities,
etc. Early Copernicus results are gathered in Spitzer, L., Jr.,
Jenkins, E.B. (1975, Ann. Rev. Astron. Astrophys, 13, pp. 133-164)
but much more is scattered in more recent literature. Observations
between 100 and 900 A will sometimes be possible and give access
to other ions.
        - Optical : Diffuse light from reflexion nebulae and inside

HII regions and interstellar extinction are extremely important
to understand IS dust (Savage and Mathis, see above). Absorption
lines are not many, but can be studied with a very high spectral
resolution, yielding important kinematical data on IS clouds and
allowing better abundance analysis (extending to the UV lines) :
Münch, in Middlehurst and Aller, op. cit., and references by
Hobbs and others in the Astrophys. J..The optical range is the
choice region to study emission lines from warm ionized ISM which
are numerous and intense : kinematics, physics and abundances in
HII regions, planetary nebulae, supernova remnants, bubbles, very
diffuse ionized gas. See the cited books by Osterbrock and Goudis.

- Infrared : For reasons of sensitivity, this spectral
region is still less interesting for the diffuse medium (with the
exception of HII regions) than for the denser molecular clouds.
Far-IR forbidden line studies of HII regions are in rapid progress and are very important complements to optical studies (see
e.g. a short review by Lacy, J.H., 1981, in Infrared Astronomy,
IAU Symposium N° 96, C.G. Wynn Williams and D.P. Cruishank, ed.,
Reidel, Dordrecht, pp. 237-245). The thermal dust emission is
reviewed e.g. by Puget J.-L., 1982, ESTEC Symposium on Submilli-
meter Astronomy, in press) and will probably be observed from
diffuse clouds in a near future.

- Radio : This is a very important domain for IS studies
which will be detailed in the next chapter. One of its interest
is that interstellar extinction is absent. Kellermann and Verschuur (ed.), Galactic and Extragalactic Radio Astronomy (book
cited above), provide most of the bases. Synchrotron continuum
emission tells us about relativistic electrons and magnetic
fields in supernova remnants and in the whole Galaxy, while
thermal continuum emission is a powerful means for studying the
ionized ISM, in particular when optical observations are made
difficult by extinction. The 21-cm line is one of the major tools
for studying IS clouds : physics, kinematics, distribution, etc.
but molecular lines (OH, CO, etc.) bring much complementary
information (the literature on this subject is unfortunately
rather scattered). Recombination lines are very interesting for
studying ionized regions but also relatively dense neutral clouds
(carbon lines): Dupree, A.K., Goldberg, L., 1970, Ann. Rev.
Astron. Astrophys., 8, pp.231-264, and Brown et al., 1978, ibid.,
16, pp.445-485.

- Magnetic field : This deserves a special mention ; the
ways to derive it are extremely varied and are summarized by
Heiles, C., 1976, Ann. Rev. Astron. Astrophys., 14, pp. 1-22.

- Cosmic rays : This important component of the ISM is
reviewed in many specialized papers. I recommend reading Cesarsky, C.A., 1980, Ann. Rev. Astron. Astrophys., 18, pp.289-319.

III. RADIO LINE OBSERVATIONS OF THE DIFFUSE INTERSTELLAR MEDIUM

1. Line excitation and transfer in the radio range

Consider a 2-level atom (or molecule) and call $n_1(V)$ and $n_2(V)$ the respective populations per unit volume of these levels, for atoms having the radial velocity V. If I_ν is the intensity of radiation per unit frequency at the frequency ν of the corresponding transition, the increase of this intensity (per unit solid angle) through crossing a length ds of the medium writes :

$$dI_\nu = h\nu \left[n_2 A_{21} + n_2 B_{21} I_\nu - n_1 B_{12} I_\nu \right] ds \qquad (1)$$

A_{21} being the Einstein probability for spontaneous emission,

$B_{21} I_\nu$ that for stimulated emission and $B_{12} I_\nu$ that for absorption. This can be identified with the transfer equation. Given the relations between these probabilities :

$$g_1 B_{12} = g_2 B_{21} \qquad (2)$$

(g_1, g_2 being the statistical weights of the levels), and

$$B_{21} = \frac{c^2}{2h\nu^3} A_{21} \qquad (3)$$

eq. (1) can be written

$$\frac{dI_\nu}{ds} = h\nu \left\{ n_2 A_{21} - \frac{c^2}{2h\nu^3} A_{21} \frac{g_2}{g_1} n_1 \left[1 - \frac{g_1 n_2}{g_2 n_1} \right] I_\nu \right\} \qquad (4)$$

(1) or (4) can be identified with the transfer equation :

$$\frac{dI_\nu}{ds} = j_\nu - K_\nu I_\nu \qquad (5)$$

hence the absorption coefficient K_ν and the emissivity j_ν .

Here comes the main difference between optical and radio lines (at least in usual interstellar conditions). In optical transitions, the population of the (high energy) upper level 2 is usually much smaller than that of the lower level 1 either because the medium is cold or (when hot) because it is dilute and collisions do not populate much the upper level. Consequently $g_1 n_2 \ll g_2 n_1$. Consequently, stimulated emission is negligible, and K_ν depends only on the population of the lower level, and usually, if this level is the fundamental one which contains essentially all atoms or molecules, of the total density of these atoms or molecules : thus absorption lines provide a direct

measurement of the column density. Conversely, the levels between
which radio transitions occur are close in energy and have
usually populations not very much different from each other, thus
the stimulated emission "correction" $g_1 n_2 / g_2 n_1$ can be very
close to unity and plays a fundamental rôle.

Radioastronomers usually introduce temperatures in order to
simplify calculations. In particular the population ratio of the
levels can be expressed in terms of an excitation temperature
T_{ex} : thanks to an extension of the Boltzman equation (valid only
at thermal equilibrium),

$$\frac{n_2}{n_1} = \frac{g_2}{g_1} \exp(-h\nu/kT) \ , \tag{6}$$

one writes :

$$\frac{n_2}{n_1} = \frac{g_2}{g_1} \exp(-h\nu/k T_{ex}) \tag{7}$$

If $h\nu \ll k T_{ex}$ (the general case for centimeter and longer
wavelengths) the absorption coefficient writes :

$$K_\nu \approx \frac{c^2}{2\nu^2} A_{21} \frac{g_2}{g_1} n_1 (V) \frac{h\nu}{kT_{ex}} \ , \tag{8}$$

thus absorption is inversely proportional to the excitation
temperature T_{ex}.

One can also express the radiation intensity I_ν in terms
of a radiation temperature $T_r(\nu)$ (which is equal to the physical
temperature of the emitter if one is dealing with black-body
radiation) :

$$I_\nu \approx \frac{2k\nu^2}{c^2} T_r (\nu) \qquad (h\nu \ll kT_r, \text{ Rayleigh-Jeans} \tag{9}$$
$$\text{approximation)}$$

If such a radiation falls on an antenna, the signal at the output
of the antenna is described by an antenna temperature $T_a (\nu)$ such
as :

$$T_a (\nu) = \eta T_r (\nu) \ , \tag{10}$$

η being the antenna efficiency (usually η is of the order of
0.5-0.7).
    The transfer equation reads simply in this case :

$$\frac{dT_r}{ds} = K(\nu) \, (T_{ex} - T_r)$$ (11)

If one looks through a uniform-T_{ex} cloud to a background of radiation temperature T_{ro} one sees by integration of the transfer equation :

$$T_r(\nu) = T_{ro}(\nu) \, \exp(-\tau_\nu) + T_{ex}(\nu) \left[1 - \exp(-\tau_\nu) \right]$$ (12)

where $\tau_\nu = \int K_\nu \, ds$ is the optical depth of the medium.

<u>Care</u> ! If the approximation $h\nu/k \ll T_{ex}$ or T_r is not valid, as is normal in the case of the short millimeter waves, the notation in temperature looses its interest. However it is the custom of radioastronomers to still express the intensities in terms of fictitious temperatures (<u>not</u> equal to the blackbody temperature if radiation is that of a blackbody) :

$$I_\nu = \frac{2k\nu^2}{c^2} \, T_r^*$$ (13)

Although one still has $T_a^* = \eta T_r^*$, the simplified equations (11) and (12) are no more valid and must be replaced by more complicated expressions (see e.g. Penzias in Balian et al., 1975, op. cit., pp.373-408).

2. The 21-cm line

This line corresponds to the transition between the two hyperfine sub-levels of the ground level of the H atom. This is a very forbidden transition. Consequently the level population is dominated by collisions and the levels are always in practice in thermal equilibrium : $T_{ex} = T$, the kinetic temperature. Since practically all H atoms are in the ground level and since $h\nu/k = 0.07$ K only and $g_2/g_1 = 3/1$, one has, if $n_H(V)$ is the density of atoms of velocity V : $n_H(V) = 4n_1(V)$, hence the numerical expression for K_ν

$$K_\nu = 5.45 \; 10^{-19} \; n_H(V)/T \quad cm^{-1} K^{-1}$$ (14)

where $n_H(V)$ is in atoms cm^{-3}/km s^{-1} and T is the kinetic temperature. From eq.(12) we see that if we observe a line emission (no background) from a cloud of uniform excitation temperature and <u>if the line is optically thin</u> ($\tau_\nu \ll 1$) the column density N_H is simply proportional to the observed line radiation temperature

$$N_H(V) = 1.835 \; 10^{18} \; T_r(\nu) \quad cm^{-2}(km/s)^{-1}$$ (15)

The problem is to know if the line is optically thin! This is obviously the case if $T_r \ll T$ (eq.12) and in general one should guess the kinetic temperature T to see whether this assumption is justified or not.

In some cases, T can be measured : this is when the 21-cm line can be observed in absorption in front of a continuum radio-source and in emission in a close-by OFF position. If T_{ro} represents the intensity coming from the radiosource at frequencies outside the line, one receives :

OFF position (emission only) :

$$T_r(\nu) = T \left[1 - \exp(-\tau_\nu)\right]$$

ON position (emission + absorbed source radiation) :

$$T_r(\nu) = T_{ro} \exp(-\tau_\nu) + T \left[1 - \exp(-\tau_\nu)\right]$$

Since T_{ro} is known from observations outside the line, one can obtain from the two measurements both the cloud temperature T and its optical depth τ_ν as a function of frequency (or velocity), provided that the cloud is uniform. The better the radiotelescope resolution, the closer the ON and OFF positions can be and the more valid this assumption is.

Emission-line surveys are very numerous ; the most striking result which emerges e.g. of the high-latitude surveys which are dealing with relatively nearby gas (see e.g. Heiles and Jenkins, 1976) is the paucity of interstellar clouds which would corres-pond to the conventional picture (isolated, more or less spherical entities) : the distribution of HI is rather characterized by arches, sheets, etc. It must be acknowledged however that 21-cm studies of HI clouds are plagued by velocity confusion problems: the half-intensity line widths are of the order of 10 km s^{-1}, due to (turbulent or bulk?) internal motions : this makes the separation of various features difficult, and also tends to wash out fine structure (if any). Indeed there are apparently no detectable emission fluctuations at scales $\lesssim 1'.5$, even in the galactic plane, and emission is practically resolved in observ-ations with interferometers. Crovisier and Dickey (1982) have shown that the power spectrum of the 21-cm fluctuations in emission is very steep (slope ≈ -3). High velocity clouds however still show fine structure in emission down to a scale of 1' (Schwarz and Oort, 1981), perhaps visible only because of their isolation (remember also that their distance is unknown and can be large) ; these features are apparently embedded in a smooth broad-lined, perhaps hotter component.
Comparison between emission and absorption 21-cm profiles at

(or near) the same location is extremely instructive. Three facts
are really striking.

i) The 21-cm emission profiles usually possess broad wings which
are not or barely visible in absorption. One might imagine that
the emission is due to many small cold clouds superimposed into
the telescope beam, but covering only a small fraction of the
solid angle so that they will usually not be seen in absorption
in the almost infinitively narrow line of sight to an extragalactic radiosource ; statistical and physical arguments due to
Lazareff (see Field in Balian et al., op. cit., pp. 467-531)
easily kill this possibility, and the broad emission wings must
be due to genuinely hot gas. The most sensitive 21-cm emission-
absorption measurements give temperatures larger than 10^3 K,
which may reach 10^4 K (Lazareff, 1975 ; Mebold and Hills, 1975 ;
Dickey et al., 1977). Although separation between hot clouds or
cloud envelopes, and a more diffuse cloud gas is obviously
difficult, these observations are the main basis for the 2-compo-
nent model of the ISM described in the first section. However
observations give only column densities, not densities, and do
not say how much of the IS volume is filled by this gas. This
filling factor can be large . From Falgarone and Lequeux (1973)
the average density of the hot ISM is of the order of $\langle n_H \rangle \approx 0.15$
cm^{-3}, with large errors. If uniform, its pressure would be
$n_H kT \approx 150 - 1500$ k $(cm^{-3}K)$ for $T = 10^3 - 10^4$ K. The upper value
is consistent with what is generally considered as the pressure
of the ISM.

ii) The well-defined features ("clouds") are always narrower in
velocity in absorption, by a sizeable factor ; the half- widths
in the absorption features have a broad distribution centred on
$4km\ s^{-1}$, corresponding to mildly supersonic motions (supersonic
turbulence or bulk motions?) given the temperature of these
clouds. Remembering that absorption is weighted towards cold
gas $(K \propto 1/T)$, one may imagine that it is due to colder, denser
(pressure equilibrium) gas sitting in clumps amongst hotter, more
diffuse gas : this reminds us of the stratification of the clouds
in the 3-component model. If a large part of the column density
is contained in these clumps at the central velocity of the
profile, one can obtain their opacities and spin temperature from
observations (if not, temperatures would be upper limits for the
colder medium). Opacities are such that optical depth effects are
clearly important in 21-cm emission surveys near the galactic plane.
Spin temperatures show a very broad distribution from 10-20 K to
$\geqslant$ 1000 K. It is not clear that the line of sight actually
crosses the cold region in the latter case ; moreover there is no
clear-cut separation between these hot clouds and the "Intercloud
medium" discussed above. One notices a tendency for temperature
to decrease with increasing optical depth (Lazareff, 1975 ;
Dickey et al., 1977 ; Crovisier, 1981 ; Payne et al., 1982a and

b ; Mebold et al., 1982). The physical meaning of this correlation is still not clear. The optical depth/temperature statistics
allow to derive the distribution of column density (normalized to
b = 90°).

$$\psi(N_H) \simeq 1.2 \ 10^{-21} \ (N_H/10^{20})^{-1.3} \quad \text{(Crovisier, 1981), or}$$

$$4.8 \ 10^{-21} \ (N_H/10^{20})^{-1.8} \quad \text{(Payne et al., 1982, Mebold}$$
$$\text{et al., 1982)}$$

With further assumptions (e.g. spherical clouds, constant pressure in the ISM) one can estimate the cloud mass spectrum, e.g.
$N(M)dM \propto M^{-1.8} \ dM$ (Crovisier, 1981). The interpretation of the
latter relation should be made with much caution, however.

iii) Although clouds observed in emission show little structure
(for reasons given above) this is not the case for absorption
observations. Fluctuations within linear distances of less than
one parsec appear to be common ; this field is presently under
active investigation by observing with interferometers absorption
in front of extended radiosources.

3. Radio molecular line observations of diffuse ISM

The physics of these lines is more complicated than that of the
21-cm line. The observed transitions are permitted, then the
radiative transitions may or may not dominate over collisional
transitions and thermal equilibrium cannot be assumed in general.
CO, which has weaker transitions, is less far from equilibrium
than OH and other molecules with strong transitions, which
exhibit complicated excitation effects.

     CO is present in diffuse clouds and is detected both by its
radio and UV transitions. It gives emission lines at 2.6 mm
($J = 1 \rightarrow 0$ transition), 1.3 mm ($J = 2 \rightarrow 1$), etc. If these lines
were unsaturated, one could derive directly the column density of
the lower level (see eqs. 8 and 12). Unfortunately the CO lines
are saturated even when they come from diffuse clouds ; one
should rather use the corresponding lines of the isotopic ^{13}CO
molecule, hoping that they are not saturated, ^{13}CO being about
50 times less abundant than ^{12}CO in diffuse clouds (see for the
direction of ζ Oph Crutcher and Watson, 1981) ; however, the ^{13}CO
lines are weak and very difficult to observe. Moreover, even if
we had the column density of the ground level J=0 of ^{13}CO, we
should also determine the excitation temperature in order to
determine the column density of the J=1 level which is likely to
be not negligible ($h\nu/k$ = 5.5 K at 2.6 mm) : this cannot be determined directly by radio observations for ^{13}CO. For ^{12}CO, if
the line is saturated, the radiation temperature in the line
should be equal to the excitation temperature (more exactly, the

intensity of the line is equal in this case to that of a black-body with the temperature equal to the excitation temperature) : there is no full guarantee however that T_{ex} is the same for ^{13}CO ; moreover the $^{12}CO/^{13}CO$ ratio has to be known if one wishes to derive the total ^{12}CO column. Thus the radio observations of CO in diffuse clouds are of limited interest, contrary to the far-UV observations of CO absorption lines which provide all required quantities (see e.g. Wannier et al., 1982). However, radio obser-vations have been used to suggest that CO is confined in the densest clumps within diffuse clouds, the CO line widths being even narrower than the 21-cm absorption line widths (Kazès and Crovisier, 1981).

As said above, the physics of OH in interstellar clouds is more complicated than that of CO ; however, there are more possi-bilities for observations in radio(the UV observations are conversely more difficult and less interesting). Emission-absorp-tion observations similar to the 21-cm ones are possible for all four lines of OH near 18 cm and have been performed in the direction of several strong continuum radiosources (see e.g. Nguyen-Q.-Rieu et al., 1976 and Crutcher, 1979). The excitation temperatures and optical depthes can thus be determined, and T_{ex} turns out to be different for each line, in particular the 1720 MHz satellite line is generally inverted ($T_{ex} < 0$). The lines are all optically thin, simplifying the interpretation. Models for excitation of these lines have been built by Guibert et al. (1978) and Bujarrabal and Nguyen-Q.-Rieu (1980) ; they are too complicated to be described here, but the interesting fact is that it is not hopeless to derive from OH lines some physical parameters for the cloud where they are formed (cf. Table 2 of Bujarrabal and Nguyen-Q.-Rieu, 1980). Like CO, OH appears from its line width to be concentrated in the densest clumps of the diffuse clouds (see Kazès and Crovisier, 1981), and perhaps even more than CO since its lines are even narrower.

References

Bujarrabal, V., Nguyen Quang Rieu, 1980, Astron. Astrophys., 91, pp.283-289
Crovisier, J., 1981, Astron. Astrophys., 94, pp.162-174
Crovisier, J., Dickey, J.M., 1982, in preparation
Crutcher, R.M., 1977, Astrophys. J., 216, pp.308-319
Crutcher, R.M., 1979, Astrophys. J., 234, pp.881-890
Crutcher, R.M., Watson, W.D., 1981, Astrophys. J., 244, pp.855-862
Dickey, J.M., Salpeter, E.E., Terzian, I., 1977, Astrophys. J. Letters, 211, pp. L77-81
Falgarone, E., Lequeux, J., 1973, Astron. Astrophys., 25, pp.253-260
Guibert, J., Elitzur, M., Nguyen-Quang-Rieu, 1978, Astron. Astro-

phys., 66, pp. 395-405

Heiles, C., Jenkins, E.B., 1976, Astron. Astrophys., 46, pp.333-360

Kazès, I., Crovisier, J., 1981, Astron. Astrophys., 101, pp. 401-408

Lazareff, B., 1975, Astron. Astrophys., 42, pp. 25-35

Mebold, U., Hills, D.L., 1975, Astron. Astrophys., 42, pp. 187-194

Mebold, U., Winnberg, A., Kalberla, P.M.W., Goss, W.M., 1982, Astron. Astrophys., in press

Nguyen-Quang-Rieu, Winnberg, A., Guibert, J., Lepine, J.R.D., Johansson, L.E.B., Goss, W.M., 1976, Astron. Astrophys., 46, pp. 413-428

Payne, H.E., Salpeter, E.E., Terzian, Y., 1982a, Astrophys. J., Suppl., 48, pp. 199-218

Payne, H.E., Salpeter, E.E., Terzian, Y., 1982b, Astrophys. J., in press

Schwarz, U.J., Oort, J.H., 1981, Astron. Astrophys., 101, pp.305-314

Wannier, P., Penzias, A.A., Jenkins, E.B., 1982, Astrophys. J., 254, pp. 100-107

ABSORPTION LINE SPECTROSCOPY OF THE INTERSTELLAR MEDIUM

Michael Jura

Department of Astronomy, UCLA, Los Angeles CA 90024 USA

Absorption line studies of the interstellar medium are
described. The discussion is in three parts. The first
describes current views of diffuse interstellar clouds, while the
second reports the results of recent extensive surveys of
interstellar regions. The final part is an outline of possible
future observations.

GENERAL BACKGROUND

Absorption line studies of the interstellar medium have
proven to be extremely powerful probes of diffuse regions where
the amount of dust extinction is not too great. (see, for
example, Spitzer and Jenkins 1975). In particular, with
absorption line spectroscopy, it has been possible to make
detailed investigations for both clouds and intercloud regions of
densities, temperatures, thermal pressures, gas-phase abundances
fractional ionizations, molecular compositions and kinematics.
With this information, it is possible to construct models for
different regions and describe pictures for the evolution of the
matter.

A brief history of optical and 21 cm observations of the
interstellar medium is given by Spitzer (1968). The most
prominent optical lines are Na I and Ca II although over the
years a number of other species have been detected. Molecules

19

J. Audouze et al. (eds.), Diffuse Matter in Galaxies, 19–34.

such as CH, CH⁺, CN and OH have also been detected. The most
comprehensively studied line of sight has been ζ Oph which has
been most closely examined both optically (Herbig 1968) and in
the ultraviolet (Morton 1975). There also exist important
surveys of particular lines in different directions in the
optical most notably by Hobbs (1969, and subsequent studies).

While ground based observations of the interstellar medium
have proven to be quite interesting, most species have their
resonance lines in the ultraviolet which therefore requires
observations above the earth's atmosphere. Although some results
from IUE have been useful, the bulk of our information about
ultraviolet interstellar lines has come from the Copernicus
satellite which was launched in 1972 and which ceased operation
in 1981. Unlike IUE, Copernicus was sensitive to wavelengths
shortward of 1200 Å where a very large number of species have
their resonance lines in this region including H_2, O VI and the
detectable lines of deuterium. Copernicus had a higher spectral
resolution than does IUE; this is important because most
interstellar lines from clouds are narrow. Finally, the
photometric accuracy of Copernicus was better than that of IUE so
it was possible to make more precise measurements of weak
interstellar features; the weak lines are the ones from which the
most information can be obtained.

Future studies of the interstellar medium are discussed
below. Here I mention that the High Resolution Spectrograph
(HRS) of the Space Telescope (ST) will prove to be an extremely
powerful tool. Also, plans are being discussed both in ESA and
NASA to build new satellites to explore the important spectral
region between 1200 Å and 900 Å where the sensitivity of the
Space Telescope will not be very good.

I. ANALYSIS OF THE DATA

Absorption line spectroscopy of the interstellar medium can
be very simple. The densities are so low that the species are
almost entirely in the ground state. As a result, stimulated
emission is unimportant for the formation of the optical and
ultraviolet absorption lines, and the interpretation of
observations to determine column densities can be staightforward.
If a line has an oscillator strength, f, and if we fully resolve
the line to measure the residual intensity at each frequency, r_ν,
in general (see Spitzer 1978) the column density of absorbing
material is given by

$$N = (\pi e^2 f/mc)^{-1} \int (-\ln r_\nu) \, d\nu$$

If the line is not resolved, we may only measure the equivalent width, W_ν, given by

$$W_\nu = \int (1-r_\nu)d\nu$$

In the limit that the line is optically thin so that $r_\nu \gg 0$, then $-\ln r_\nu \cong (1 - r_\nu)$ and the equivalent width is directly proportional to the column density of material. Otherwise, unless the line lies on the damping portion of the curve of growth, the derivation of the column density from the equivalent width is subject to considerable ambiguity. In general, the only species whose absorption lines lie without doubt on the damping portion of the curve of growth are H and H_2.

An observational question always is whether a relatively weak line lies on or near the linear portion of the curve of growth. With infinitely good resolution it is always possible to tell by examining the line profile. In clouds where the temperature is cold (< 100 K), the thermal broadening of a relatively heavy atom may be 0.2 km s^{-1} which is much smaller than the instrumental resolution normally available. As a result, the physical uncertainty is whether "turbulent" broadening is important.

One approach to determine the minimum line width in an absorption line region is to search for the hyperfine structure in the Na I D line at a velocity separation of 1.0 km s^{-1} (Hobbs 1969). While a few clouds do display this splitting (Blades, Wynne-Jones and Wayte 1980) most do not, with the consequence that we may often assume that lines are typically broader than 0.6 km s^{-1}. Another approach to measure of the velocity dispersion in clouds is to make radio observations of molecular emission from the diffuse clouds. Radio instrumental techniques are such that very high resolution is available. For diffuse clouds studied in absorption, the two molecules that have been used are CO and CH. Again, while some lines of sight display very narrow features (Liszt 1979), many show "turbulent" broadening on the order of 2-3 km s^{-1} (Knapp and Jura 1976, Lang and Willson 1978). A final way to determine the intrinsic line widths is to use curve-of-growth techniques (see Spitzer 1978). This procedure is limited if, for example, there are two clouds in the line of sight of which one has a very narrow intrinsic line width (see Nachman and Hobbs 1973).

Given the ambiguities in interpreting saturated lines; for many purposes it is clear that it is essential to study weak lines or lines that are so strong that they are on the damping portion of the curve-of-growth.

II.   BASIC RESULTS

   A.   Distribution and kinematics

   One of the most important results of the extensive studies
of the interstellar absorption lines is that the gas is very
clumpy. Even stars that are less than 100 pc from the sun
characteristically show several different interstellar components
(e. g. York and Kinahan 1979). At the moment, the only
comprehensive surveys of matter in different components have been
of species such as Na I which is only a minor fraction of all of
the sodium; hence it is difficult to determine the total amount
of material in different clouds (Hobbs 1974). As a result, the
mass spectrum of diffuse interstellar clouds is not well known.

   Classically (e. g. Spitzer 1978), it is known that most
diffuse clouds have velocities < 20 km s^{-1} in the LSR. These
clouds contain the bulk of the mass and essentially all the
molecules. However, there are some high velocity clouds as well
which presumably are the result of various explosive processes in
the interstellar medium (Spitzer and Jenkins 1975, McCray and
Snow 1979).

   B. Abundances

   Abundances in clouds have been extensively measured. One of
the major results (discussed here by de Jong and Spitzer and
Jenkins 1975) is that for $E(B-V) > 0.10$ mag, much of the hydrogen
is molecular; otherwise it is mostly atomic. In diffuse clouds,
except for hydrogen, most of the gas is in atomic form although
there are trace amounts of CO, CH and other simple diatomic
molecules. Large amounts of molecules are only found in dark
clouds where the ultraviolet does not easily penetrate.

   For species other than hydrogen, the interpretation is
complicated by the depletion onto solid grains. Abundances of
some volatile elements such as oxygen are generally not more than
factor of 2 below the solar abundance (de Boer 1981). There is
no evidence for any overabundances. Refractory elements are
substantially depleted; in some cases by more than a factor of
100. The usual interpretation of this result is that the
refractory elements are largely contained within interstellar
grains while the volatiles are not (see Field 1974). However,
the explanation for this difference beween volatiles and
refractories is not clear. While grain cores form in the
atmospheres of cool stars, it also seems possible that the
observed pattern of depletion could occur in the interstellar
medium itself (Snow 1975, Jura 1980).

C. Temperatures, Densities and Pressures

Cloud temperatures have been best derived from analysis of the H_2 absorption lines. When there is a large amount of H_2, the ratio of the amount in $J = 0$ compared to the amount in $J = 1$ is a good measure of the cloud temperature (Spitzer and Cochran and Hirshfeld 1974), and most clouds have a value near 80 K. For clouds with relatively little H_2, the determination of the temperature is much more difficult. It is usually assumed that T = 80 K, but this need not in general be correct (see Dickey, Salpeter and Terzian 1977).

Densities and the resulting pressures are derived from analysis of H_2 (Jura 1975a,b) and the fine structure population of C I (see below). While a range of densities are found, typical values range between 10 and 100 cm^{-3} with some values as high as 1000 cm^{-3}. The thermal pressures corresponding range from 1000 to 10,000 cm^{-3} K although some higher and lower pressures are found.

D. Intercloud Medium

This matter is mostly discussed here by Boksenberg. The point of the interstellar studies is that a wide range of intercloud properties are found with ample evidence for gas with temperatures up to 10^6 K (Spitzer and Jenkins 1975).

RESULTS FROM RECENT ULTRAVIOLET SURVEYS

Before the operation of the Copernicus satellite was terminated, considerable time was devoted to two extensive surveys; the results of which are now being published. Here I discuss these relatively recent results.

I. THE ABUNDANCE SURVEY

One of the most important early results of the Copernicus observations was the discovery that substantial depletion of some elements occurs within interstellar clouds. This is of course consistent with the expectation that the grains must contain at least 1% of the mass of the gas (Spitzer 1978, Aannestad and Purcell 1973). The next step is to be much more quantitative about these observations. For example, how sensitive are they to density or to the charge or mass of the atom that is depleted. How do depletions correlate with the amount of dust?.

Qualitatively, one of the first results of the Copernicus survey was that "refractory" elements were the most depleted while volatile elements were much less so (Field 1974). A standard interpretation of this result is that grains are formed in a region of 1000 K, for example in circumstellar shells of late type stars. Those substances which form solids at these temperatures are confined within the grains, while those atoms which remain in the gas phase do not get trapped into the solids. However, it may also be that the observed depletions reflect the chemical properties of the atoms when they collide with grains in the interstellar medium rather than thermodynamic properties in cirucumstellar envelopes of late type stars. For example phosphorus is apparently undepleted (Jura and York 1978) even though it seems to have a condensation temperature close to that of iron which is depleted by typically a value near 100 (Savage and Bohlin 1978).

Quantitatively, the early Copernicus results were not completely definitive. For example, Morton (1974, 1975) derived a depletion of oxygen toward the well studied star ζ Oph of about a factor of 5 below the solar value. More recently, de Boer (1981) has disputed this result and has argued that oxygen is not depleted by more than about a factor of 1.3.

In order to address these questions, Bohlin et al. (1982) have performed a systematic survey of ultraviolet absorption lines in different clouds. They observed 88 stars and they measured primarily weak lines for the reasons described above. Suitable weak lines are available for O I, Si II, P II, S II, Cl I, Cl II, Mn II, Fe II, Ni II and Cu II. In their first paper the data are presented; they defer until later paper a scientific analysis of their results.

However, even without sophisticated interpretation of the data, some preliminary results can be obtained. In Table 1, I show the oxygen abundances derived from a simple analysis. I take the column density of hydrogen nuclei from Bohlin, Savage and Drake (1978) and the oxygen column density from the measurements of Bohlin et al. (1982) of the weak O I 1355 Å line which has an f value of $1.25 \ 10^{-6}$ (Morton 1978). No attempt is made to correct for saturation in the O I line. In Table 1, I list in order of Right Ascension only those stars for which the O I line was detected; most of the 88 stars were either too faint or had too little matter in the line of sight for the O I line to be detected.

Table 1 - Interstellar Oxygen Abundances

Star	$[O]/[H] \cdot 10^4$
40 Per	3.7
o Per	3.6
ε Per	5.3
ξ Per	2.4
α Cam	6.2
ϕ^1 Ori	7.5
λ Ori	4.7
κ Ori	4.5
1 Sco	4.1
δ Sco	3.1
β^1 Sco	2.8
ω^1 Sco	3.4
ν Sco	4.9
σ Sco	3.0
ρ Oph A	1.4
χ Oph	5.7

The important conclusion to be derived from Table 1 is that compared to the solar abundance of $7 \cdot 10^{-4}$ (Morton 1974), oxygen is always somewhat but not greatly depleted. Considering that we make no correction for saturation, the abundances could even be slightly higher than we show. These results are consistent with the model that the grains contain moderate amounts of silicates (Mathis, Rumpl and Nordsieck 1977), but there is no reason to think that the grains contain substantial amounts of ice or other oxygen bearing molecules. It will be most interesting to continue this analysis for the other species measured by Bohlin et al. (1982) and to determine if the data are of sufficiently high quality to see if there are correlations or anti-correlations between, say, the oxygen depletion and the amount of dust.

An important scientific point follows. We would expect that for a density of 100 cm^{-3}, a temperature of 80 K and a grain to gas cross section of 10^{-21} cm^2 (see Spitzer 1978) that the collision time of an oxygen atom with a grain is $3 \cdot 10^{14}$ s (10^7 years). This is comparable to the lifetime of a typical cloud. Since most of the clouds, except possibly the one toward ρ Oph, display relatively little depletion, it seems that oxygen atoms do not adhere to grains for a very long time. Quantitatively, in particular, the lifetime of an oxygen atom (and the resulting molecules that it may form) on the surface of a grain is probably less than 10^6 years.

B. The CI SURVEY

A critical parameter for understanding the dynamical
behavior of the interstellar medium is the thermal pressure of
the gas. The measurement of this quantity requires an estimate
of the density and temperature. As mentioned above, it appears
that the range of temperatures in diffuse clouds is not too
great, but that the range of densities is substantial. It turns
out that one of the most suitable ways to study densities and
temperatures in diffuse clouds is to observe the populations in
the ground state fine structure lines of this species (see, for
example, Jenkins and Shaya 1979). (Incidentally, infrared
emission from these lines has been detected by Phillips et al.
(1980) in dark molecular clouds.) It can be shown that in most
cases optical pumping of the C I fine structure population is
small and that the level populations are controlled by the
collisions with other atoms.

C I is a particularly useful probe of the densities in
diffuse clouds for several reasons. First, C I has a rich
ultraviolet spectrum with a wide range of f values; therfore, it
is possible to determine accurate column densities. Also, C I is
abundant in many different regions becuse there is so much carbon
in the interstellar medium. Finally, the populations of the fine
structure levels are sensitive to the physical conditions that
prevail in diffuse interstellar clouds.

A preliminary set of C I data were obtained by Jenkins and
Shaya (1979) from the original Copernicus survey. These
observations were not optimized for the determination of gas
pressures, and so we (Jenkins, Jura and Loewenstein 1982)
undertook a more comprehensive survey with Copernicus to develop
more accurate observations of the interstellar pressure.
Unfortunately, because of the decline in sensitivity of
Copernicus, our results were not as large an improvement over the
old data as we had hoped.

In this more recent survey, 27 stars were observed, and C I
column densities or upper limits were accurately derived for most
of the different lines of sight. As is often the case, with such
a large body of data, it was necessary to argue that some of the
f values of the C I lines should be revised from their previous
values.

It generally was possible to make estimates for the thermal
pressure of the gas averaged over the different clouds in the
entire line of sight to the observed star. In Table 2 we show
our results and compare them with the densities derived from
analysis of the H_2 lines as well.

Table 2 – Pressures ($\log[p/k]$ in units of cm^{-3} s^{-1})

Star	lower limit	upper limit	H_2 result
κ Cas	2.5	3.5	>3.6
γ Cas	--	3.3	<3.0
40 Per	3.7	3.7	
ε Per	2.65	3.0	3.0
ξ Per			4.4
(90%)	2.7	3.25	
(10%)	4.6	--	
α Cam	--	2.5	4.1
δ Ori A	--	3.1	2.9–3.4
λ Ori A	3.2	3.7	3.3
ι Ori	--	3.3	2.9–3.4
ε Ori	2.9	3.55	
κ Ori	3.0	3.5	
139 Tau	--	3.4	3.5
HD 74375	3.1	3.95	
1 Sco	3.7	3.8	>3.4
β¹ Sco			>3.3
(70%)	--	3.4	
(30%)	4.7	--	
ω¹ Sco	3.35	4.15	
ν Sco	3.8	4.5	
ρ Oph A	2.9	4.3	
χ Oph	--	4.7	>3.6
22 Sco	2.6	4.7	
μ Nor	--	3.45	4.4
γ Ara	3.0	3.5	3.3
θ Ara	3.0	3.5	<2.7
κ Aql	2.9	3.55	<3.4
59 Cyg	3.05	3.75	3.85
σ Cas	3.1	3.5	

The results in Table 2 show several important results.
First, the pressures do seem to be in the range between 10^3 and
10^4 cm^{-3} s^{-1}, and this is a distinct contraint on any model for
these clouds (see Lazareff, these lectures). Also, there are
clouds both with unusually low and unusually high pressures. The
interstellar medium is not isobaric, and it is subject to
considerable pressure variations, presumably because of a variety
of dynamical processes that occur. Finally, while the H_2 and C I
analysis usually agree moderately well, there are conspicuous
problems; the clearest case is toward α Cam. We have carefully
checked our C I results toward this star, and it seems that the
pressure is almost certainly as low as implied by this result.
However, the pressure derived from analysis of the H_2 (Jura
1975b) assumed that the formation and destruction rates of this
molecule are in a steady state. It is known that α Cam is a

runaway star, and it is likely to be rapidly approaching some of
the interstellar matter in its direction. As a result, in a
steady state, there should be appreciably more atomic hydrogen
that is currently found, and as a consequence, the density is
much lower than derived by this analysis. Therefore, the
pressure toward α Cam is probably best measured by the C I data,
and it seems clear that the pressure in this direction is much
lower than predicted by some theoretical models of the
interstellar medium.

POSSIBLE FUTURE OBSERVATIONS

In the next few years, we will have an exceptional
opportunity to improve dramatically our understanding of the
interstellar medium with the High Resolution Spectrograph on the
Space Telescope. This instrument will have a number of gratings
with resolutions between 10^3 and 10^5 which are well suited for a
number of studies. Furthermore, it will have a Digicon detector
which should prove to be both sensitive and linear so that it
should be possible to determine accurate line profiles and
equivalent widths of even very weak lines. The HRS will be most
sensitive for wavelengths > 1200 Å, and it should still be useful
for wavelengths > 1100 Å. This extra 100 Å of moderate
sensitivity should prove very useful in studying H_2 which has its
(0,0) Lyman band at 1108 Å.

There are at least two additional satellites under study to
perform observations that the HRS will not be able to do in the
region shortward of 1200 Å. ESA is considering Magellan while
NASA is considering FUSE (Far Ultraviolet Spectroscopic Explorer.
Neither satellite has received funding and so it will be a number
of years before either is operating, but at least it should be
remembered that these satellites may provide information that we
may be unable to acquire with the Space Telescope.

Because the HRS will be operational in the next few years,
here I present a list of possible observations that may be
performed with this instrument. This list of projects is
subjectively chosen, but I hope it reflects the possibilities
that will soon be available.

I. C II Abundances

Oxygen and nitrogen gas phase abundances have been
reasonably well determined by the recent Copernicus survey of
Bohlin et al. (1982) described above. However, the C II
abundance has not been well measured because the lines that
Copernicus could study were very heavily saturated. All that can
be said with confidence is that carbon is not depleted by more

than a factor of 10. There is, however, a very weak line of C II
at 2325 Å; this line has an f value of 6.7×10^{-8} (Cowan, York and
Hobbs 1982) and it has been detected with <u>Copernicus</u> toward δ Sco
(Hobbs, York and Oegerle 1982). A systematic survey with the HRS
will be possible to determine the typical carbon depletion.

This question of the carbon depletion is critical for
several problems in understanding diffuse clouds. First,
excitation of the fine structure line of C II with an excitaion
temperature of 92 K above the ground state is the dominant
cooling process in diffuse clouds (Dalgarno and McCray 1972, Jura
1977). At the moment, it seems likely that a cooling rate of
10^{-25} erg H nuclei s^{-1} is common within diffuse clouds, but this
should be confirmed in more detail by observations. Such a high
cooling rate is important because it drastically restricts
possible models for the heating of the diffuse clouds;
undoubtedly the best candidate to date is the photoheating off
grains.

The gas phase abundance of carbon is important for
understanding the nature of the grains in the interstellar
medium. Presumably carbon stars eject carbon grains into the
interstellar medium; it would be most interesting to discover if
the amount of carbon in the gas is consistent with much of the
carbon being in grains. For example, it is not clear whether the
2200 Å feature is carried by graphite as is usually believed or
by some other substance.

The gas phase abundance of carbon in diffuse clouds is
related to an unsolved problem for dark clouds, where is all the
carbon? It is often argued that most of the carbon ought to be
in the form of CO. However, it appears that only about 10% of
the solar abundance of carbon is in this molecule (Dickman 1978).
Is this because carbon is generally depleted onto grains? If so,
we might expect a similar depletion to occur in diffuse clouds.

Finally, it seems very important for our understanding of
the process of nucleosynthesis in the galaxy to determine the
abundance fluctuations of carbon, nitrogen and oxygen relative to
each other in diffuse clouds. The measurements of the true
abundance of these species is complicated by the possiblity that
there may be significant depletion of some of the material onto
grains --empirical studies will help determine this question. In
any case, cloud by cloud investigations with the HRS should make
it possible to see how this quantity varies in different regions,
and as a function of cloud density.

B. C I Survey

As discussed above, C I provides an extremely valuable probe
of cloud pressures. However, Copernicus observations were
unable to resolve individual components (or different clouds) in
different lines of sight. With the HRS, it should be possible to
do this. Several consequences follow.

(i) Interstellar Shocks

Spitzer and Morton (1976) have used H_2 profiles from Copernicus
to identtify expanding shocks around H II regions. This process
has been studied theoretically for many years, but it has not
been firmly established observationally. The Copernicus results
certainly seemed to indicate that there were expanding shocks
around, bright, early type stars, but it would be most important
to confirm these observations with studies of C I.

Another set of shocks that has been identified are the "high
velocity" clouds -clouds with velocities greater than 20 km s^{-1}
in the LSR. Some of these clouds clearly contain C I absorption
lines (for example, toward ρ Leo), and it would be most worthwhile
to obtain high resolution measurements of these clouds. It
should be possible in cases of favorable geometry to measure the
density as a function of velocity and thus to compare with models
of these regions.

(ii) Individual cloud pressures

As described above, C I is extremely useful for determining
cloud pressures, but the current observations do not allow cloud
by cloud studies of this quantity. With the HRS, it should be
possible to measure abundances in individual clouds. This will
be most imporant for determing the amount of variation in the
thermal pressure in different regions in the Galaxy, and for
understanding depletions in different clouds.

C. Dust to Gas Ratio in Dark Clouds

One of the most important results from ultraviolet
observations of the interstellar medium is the result that in
diffuse clouds within about 1 kpc of the sun, the dust to gas
ratio is relatively uniform (Bohlin, Savage and Drake 1978).
However, this ratio has only been studied in diffuse clouds with
E(B-V) $<$ 0.35 mag. With the low resolution mode of the HRS
(10^3), it should be possible to measure the amount of H_2 in the J
= 0 and J = 1 rotational bands toward more heavily reddened
stars., Also, we will be able to determine the column density of
atomic hydrogen from the profile of the Lyman α line at 1216 Å.

From these studies, it should be possible to determine the total
gas to dust ratio in clouds with E(B-V) < 1 mag. This is
particularly interesting because such clouds are self-gravitating
and on the verge of collapsing and forming stars. As a result,
it is important to establish the dust to gas ratio in these
clouds which are qualitatively different than the clouds
currently studied. It may be possible to determine whether
substantial dust evolution occurs in these clouds of generally
higher density.

D. Detailed Studies of CO

     CO is one of the most important interstellar molecules.
Studies with the HRS should greatly improve our understanding of
this species. Several specific studies come to mind.

     (i) Abundances

     First, it should be possible to measure accurately the
amount of this species in different clouds. Federman et al.
(1980) used Copernicus - data to study CO vs. C I. One would
expect a correlation between these two species because they are
both formed as the density squared and destroyed by the
absorption of ultraviolet. However, there is considerable
uncertainty in the quantitative analysis by Federman et al.
(1980) because of the difficulty of measuring accurately the
equivalent widths of very weak lines with Copernicus.
The HRS will provide much better data.

     (ii) Chemistry

     It should be posible to determine whether current models for
the formation of CO are reasonable. That is, there is a
tentative Coperncius detection of CO toward ι Ori, a star with
very little H_2. It should be possible to search for CO toward
this star and determine if it is present. If it is, this would
imply that a substantial amount of CO can be formed on grains.
If on the other, hand, CO is only found in regions with abundant
amount of H_2, this implies that the usual gas-phase scheme for
the formation of this species are dominant.

     It may be possible to understand better the problem of the
photodestruction of CO. At the moment, good laboratory data are
not available. Apparently, however, CO is mostly destroyed by
absorption in lines rather than in a continuum (Bally and Langer
1982, Morris and Jura 1982). As a result, CO should become
self-shielding once its column density is high enough. This
phenomena is known to occur for H_2 (Spitzer and Jenkins 1975),
and it should be possible to determine if it occurs for CO also.

(iii) Isotopes

Another experiment will be to measure the amount of ^{13}CO vs. ^{12}CO. This has been done with Copernicus (Wannier, Penzias and Jenkins 1982), but the lack of instrumental sensitivity made the measurements very difficult. It should be much simpler to do this with the HRS.

(iv) Rotational Excitation

With the HRS it should be possible to study the populations of CO in its different rotational levels. The HRS should resolve many of these lines, and the populations are sensitive to models for excitation and the physical conditions in the clouds. This sort of analysis should be particularly interesting to radio astronomers because they use models for rotational line excitaion to explain their emission measurements. However, there are currently no direct checks of these models, and it should be possible to do this with the HRS.

E. Studies of OH and H_2O

As with CO, the study of these molecules should be possible with the HRS. Most of the Copernicus observations were not of high quality because the lines were too weak (Smith and Snow 1979, Snow and Smith 1981, Smith et al. 1981). Future observations will help clarify the oxygen chemistry in diffuse clouds. Does oxygen form molecules on the surfaces of grains, or, as expected, do gas-phase processes dominate? (see Crutcher and Watson 1976).

F. Other Studies

Here are only listed a few possible interstellar studies with the HRS; this is only the beginning of the possibilities.

REFERENCES

Bahcall, J. N., and O'Dell, C. R. 1979, in IAU Colloquium No. 54, Scientific Results with the Space Telescope, 5.

Bally, J., and Langer, W. D. 1982, Ap. J., 255,, 143.

Blades, J. C., Wynne-Jones, I., and Wayte, R. C. 1980, M. N. R. A. S., 193, 849.

Bohlin, R. C., Savage, B. D., and Drake, J. F. 1978, Ap. J., 224, 132.

Bohlin, R. C., Hill, J. K., Jenkins, E. B., Savage, B. D., Snow, T. P., Spitzer, L., and York, D. G. 1982, Ap. J. Suppl., in press.

Cowan, R. D., Hobbs, L. M., and York, D. G. 1982, Ap. J., 257, 373.

Crutcher, R. M., and Watson, W. D. 1976, Ap. J., 209, 778.

Dalgarno, A., and McCray, R. M. 1972, Ann. Rev. Astr. and Ap., 10, 375.

de Boer, K. S. 1981, Ap. J., 244, 848.

Dickey, J. M., Salpeter, E. E., and Terzian, Y. 1977, Ap. J. (Letters), 211, L77.

Dickman, R. L. 1978, Ap. J. Suppl., 37, 407.

Federman, S. R., Glassgold, A. E., Jenkins, E. B., and Shaya, E. J. 1980, Ap. J., 242, 545.

Field, G. B. 1974, Ap. J., 187, 453.

Herbig, G. H. 1968, Zs. f. Ap., 68, 243.

Hobbs, L. M. 1969, Ap. J., 157, 135.

Hobbs, L. M. 1974, Ap. J., 191, 395.

Hobbs, L. M., York, D. G., and Oegerle, W. 1982, Ap. J. (Letters), 252, L21.

Jenkins, E. B., and Shaya, E. J. 1979, Ap. J., 231, 55.

Jenkins, E. B., Jura, M., and Loewenstein, M. 1982, in preparation.

Jura, M. 1975a, Ap. J., 197, 575.

Jura, M. 1975b, Ap. J., 197, 581.

Jura, M. 1978, in Protostars and Planets, ed. T. Gehrels (University of Arizona: Tucson), 165.

Jura, M. 1980, in Highlights of Astronomy, 5, 293.

Jura, M., and York, D. G. 1978, Ap. J., 219, 861.

Knapp, G. R., and Jura, M. 1976, Ap. J., 209, 782.

Lang, K. R., and Willson, R. F. 1978, Ap. J., 224, 125.

Liszt, H. S. 1979, Ap. J. (Letters), 233, L147.

McCray, R. M., and Snow, T. P. 1979, Ann. Rev. Astr. and Ap., 17, 213.

Morris, M., and Jura, M. 1982, Ap. J., in press.

Morton, D. C. 1974, Ap. J. (Letters), 193, L35.

Morton, D. C. 1975, Ap. J., 197, 85.

Morton, D. C. 1978, Ap. J., 222, 863.

Nachman, P., and Hobbs, L. M. 1973, Ap. J., 182, 481.

Phillips, T. G., Huggins, P. J., Kuiper, T. B. H., and Miller,
    A. E. 1980, Ap. J. (Letters), 238, L103.
Savage, B. D., and Bohlin, R. C. 1979, Ap. J., 229, 136.
Smith, P. L., Yoshino, K., Griesinger, H. E., and Black, J. H.
    1981, Ap. J., 250, 166.
Smith, W. H., and Snow, T. P. 1979, Ap. J., 228, 435.
Snow, T. P. 1975, Ap. J. (Letters), 202, L87.
Snow, T. P., and Smith, W. H. 1981, Ap. J., 250, 163.
Spitzer, L. 1968, Diffuse Matter in Space (J. Wiley: New York).
Spitzer, L. 1978, Physical Processes in the Interstellar Medium
    (J. Wiley: New York).
Spitzer, L., Cochran, W. D., and Hirshfeld, A. 1974, Ap. J.
    Suppl., 28, 373.
Spitzer, L., and Jenkins, E. B. 1975, Ann. Rev. Astr. and Ap.,
    13, 133.
Spitzer, L., and Morton, W. 1976, Ap. J., 204, 731.
Wannier, P. G., Penzias, A. A., and Jenkins, E. B. 1982, Ap. J.,
    254, 100.
York, D. G., and Kinahan, B. F. 1979, Ap. J., 228, 127.

STELLAR WINDS AND THEIR INTERACTION
WITH THE INTERSTELLAR MEDIUM

Henny J.G.L.M. Lamers

Astronomical Institute
Space Research Laboratory
Beneluxlaan 21, 3527 HS Utrecht
The Netherlands

I. INTRODUCTION

The most massive stars are losing mass in the form of a more or
less steady stellar wind, which blows during most of their lifetime. The energy and momentum of this stellar wind is so large
that the wind blows a bubble into the interstellar medium (ISM).
The amount of energy which is deposited into the ISM by the stellar wind during the lifetime of the star is comparable with the
energy released in Supernova explosions. In this course I will
first review the basic knowledge of the stellar winds (Section
II). The theory of the interaction between the wind and the ISM
and the structure of a typical interstellar bubble (IB) from a
single star is discussed in Section III. A typical example of an
observed bubble, the Cygnus Superbubble, is discussed in Section
IV.

II. BASIC KNOWLEDGE OF STELLAR WINDS

The first clear evidence of stellar winds from hot stars was the
discovery of strongly Doppler-shifted P Cygni line profiles in
the UV spectrum of the Orion Belt supergiants (Morton, 1968).
Later studies of large numbers of stars by means of the Copernicus satellite and the International Ultraviolet Explorer have
contributed to an enormous extension of our knowledge. I will review the important facts: (see reviews by Conti, 1978; Cassinelli,
1979; Lamers, 1982)
    a. All stars with luminosity $L > 2.10^4$ $L_\odot$ or $M \gtrsim 12$ $M_\odot$ lose
mass during their entire lifetime. The less massive stars suffer
mass loss only in their final evolutionary stage when they are
red giants or red supergiants. Less massive stars can also lose

J. Audouze et al. (eds.), Diffuse Matter in Galaxies, 35–44.
Copyright © 1983 by D. Reidel Publishing Company.

mass during their main sequence lifetime when their rotational
velocity is large (v sin i $\gtrsim$ 300 km/s).

   b. The wind is accelerated from a small velocity (v $\sim$ 10
km/s) at the photosphere to a high velocity (v $\sim$ 10^3 km/s) within a distance of three to ten stellar radii. The terminal velocity v_∞ reached by the wind at large distances ($\sim$ 10^2 R_*) is related to the escape velocity, v_{esc}, at the stellar surface:

$$v_\infty = A \ v_{esc} \tag{1}$$

with

$$v_{esc} = (2 \ GM \ (1 - \Gamma) \ / \ R_*)^{\frac{1}{2}} \tag{2}$$

The factor $1 - \Gamma$ which is usually 0.7 to 0.9 is a correction due
to the radiation pressure by electron-scattering in the photosphere. The proportionality factor A depends on the stellar temperature and is 2.5 for O-stars, 3.5 for early B-stars, and decreases to 1.0 for A stars.

   c. The mass loss rate depends mainly on the stellar luminosity, L. Garmany et al. (1981) and Abbott et al. (1981) find that
the mass loss rate ($\dot{M}$ in $M_\odot$/yr) depends on L only

$$\dot{M} = 8.9 \ . \ 10^{-6} \ (L \ / \ 10^6 \ L_\odot)^{1.56} \tag{3}$$

However, Lamers (1981) and Chiosi (1981) have shown that the mass
loss rate also depends on the gravity

$$\dot{M} = 1.5 \ 10^{-5} \ (L \ / \ 10^6 \ L_\odot)^{1.42} \ (g \ / \ 10^3)^{-0.31} \tag{4}$$

Both relations predict about the same mass loss rates for OB supergiants, but the European formula (4) predicts a smaller mass
loss rate for the main sequence stars than the American formula
(3). Since the stars spend about 90% of their lifetime as a main
sequence star when the gravity is about a factor 3 to 10 larger
than the supergiant phase, the smaller mass loss rates found by
Lamers for the main-sequence stars imply that the total amount of
mass and energy lost by the star during its lifetime is about 1.5
to 2 times smaller than predicted by relation (3).

   d. The Wolf-Rayet stars, i.e. evolved luminous stars which
have lost most of their hydrogen layers, are an exception. Their
mass loss rates are about constant at $\dot{M} = 3 \ . \ 10^{-5} \ M_\odot$/yr , independent of luminosity, (Willis, 1982).

   In Table 1 I have summarized the basic data for stars of
different masses. The luminosity and lifetimes are from evolutionary calculations with mass loss (de Loore et al., 1979). The
mass loss rates are from equation (3). A mean wind velocity of
2000 km/s during the entire lifetime was adopted. Notice that the
values of the total mass and total energy are reduced by a factor
1.5 to 2 if equation (4) was used.

TABLE 1. MASS AND ENERGY LOST BY STELLAR WINDS

Initial mass ($M_\odot$)	L ($10^6 L_\odot$)	$\dot{M}$ $\left[\dfrac{10^{-6}}{M_\odot/yr}\right]$	v_∞ (km/s)	lifetime (10^6yrs)	total mass lost ($M_\odot$)	wind energy (10^{36}erg/s)	total wind energy (10^{51}erg)
100	1.00	5.5	2000	2.6	14	6.9	56
60	0.47	1.5	2000	3.3	5	1.9	19
40	0.21	0.34	2000	4.1	1.4	0.42	5.3
30	0.11	0.11	2000	4.8	0.5	0.14	2.1
20	0.04	0.02	2000	6.7	0.1	0.025	0.5

III. THEORY OF THE INTERSTELLAR BUBBLES

The wind which has a velocity of about 2000 km/s and a kinetic
energy of the order of 10^{36} erg/s blows a bubble in the ISM. I
will describe the structure and evolution of the IB, following the
theory of Castor et al. (1975) and Weaver et al. (1977). We assume that the wind is turned-on instantaneously when the star
starts its hydrogen burning. We also assume that the ISM around
the star is homogeneous with a density ρ_0 (g/cm^3) or n_0 (atoms/
cm^3).

a. First phase: free expansion of the wind

The wind expands freely into the ISM, almost unobstructed by
the presence of the ISM. This phase lasts until the wind has swept
up a sufficiently large mass of IS gas to stop the free expansion.
This phase will last about 10^2 yrs at which time the radius of the
bubble is 0.2 pc.

b. Second phase: adiabatic expansion of the bubble

When sufficient gas is swept-up to stop the free expansion
of the wind, either by Coulomb-stopping or by a collisionless
shock, the bubble complex will have the following structure (Figure 1). From inside to outside:
1. the supersonic wind in free expansion with radius R_1
2. a region consisting of shocked wind material with radius R_2.
   All the kinetic energy of the wind is dumped into this region
   in the form of heating, creating a temperature of about
   10^6 - 10^7K. All the stellar wind mass is in this region, so
   $M_2 = \dot{M}t$, if $\dot{M}$ is the mass loss rate of the star. Because this
   region is hot, it expands into the ISM with a velocity $v_2 =$
   d R_2 / dt.
3. Surrounding region 2 there is a shell of swept-up IS gas due
   do the expansion of region 2. This shell of swept-up gas (re-

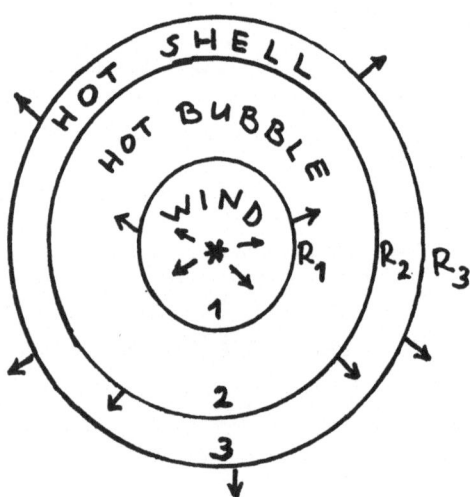

*Figure 1. The structure of the bubble-complex during the adia-
batic expansion phase.*

gion 3) has an outer radius R_3 and contains all the swept-up
IS gas. So its mass is $M_3 = (4/3) \pi R_3^3 \rho_0$. The boundary between region 3 (the shell) and the ISM is a shock due to the
expansion of the shell into the ISM. During this phase b the
expansion velocity $v_3 = d R_3 / dt$ is so large that the strong
shock heats region 3 to $T > 10^6$ K. So radiative cooling of the
shell is negligible and as a result this shell as well as the
hot bubble (region 2) will both expand adiabatically. That is
why this phase is called the adiabatic phase.
During the adiabatic phase the expansion of regions 2 and 3 are
proportional to $t^{3/5}$, with $R_3 = 1.16 R_2$, so their velocity decreases at $t^{-2/5}$. The calculations show that from the wind energy
$L_w t$ (where L_w is the kinetic luminosity of the wind, $L_w = 0.5 \ \dot{M} v_\infty^2$)
dumped into region 2 about 45% is used for heating of region 2 and
the remaining 55% has been transferred by the expansion to region
3, where 22% is used for kinetic energy of the shell and 33% for
heating of the shell.
    This adiabatic phase will end when the shell starts to cool
radiatively. This occurs when the age of the system becomes comparable to the timescale for radiative cooling, which is after a
time

$$t \sim 1.7 \quad 10^3 \quad \sqrt{10^{-36} L_w / n_0} \tag{5}$$

years. The wind luminosity of a typical O7 star is 10^{36} erg/s and
since $n_0 \sim 1$ the phase lasts only a few thousand years. At that
time the bubble complex has an outer radius $R_3 \approx 0.6$ pc. This im-

plies that there is only a small probability to observe a bubble
in the adiabatic phase. From the observational point of view the
next phase is much more interesting.

c. Third phase: Snow-plough phase of hot bubble and cold shell

When the temperature of the shell of swept-up IS gas has decreased below 10^6 K radiative cooling becomes important and the
shell will rapidly cool down to 10^4 K. The boundary between the
shell and the ISM is an isothermal shock. Since isothermal shocks
have a high compression rate, the density in the cold shell will
be high (n $\sim$ 10^1-10^3 cm^{-3}) and as a consequence the shell will be
thin. The structure of the bubble complex during this snow-plough
phase is shown in Figure 2. The hot bubble (region 2) is still too
hot to suffer radiative cooling and so it will continue to expand
adiabatically. The bubble is almost isobaric since the sound-speed
crossing time R_2 / C_s is much smaller than the age of the system.
At the boundary between the hot bubble and the cold shell the temperature gradient is so steep that thermal conduction by electrons
will transport heat from (2) tot (3). As a result of this the
heated bottom layers of the cold shell will start to evaporate
and gas will flow from (3) to (2). This has no noticable effect
on the mass of the shell, but it can contribute significantly to
the mass in region (2). We will derive the equations which de-
cribe the structure and evolution during this phase:

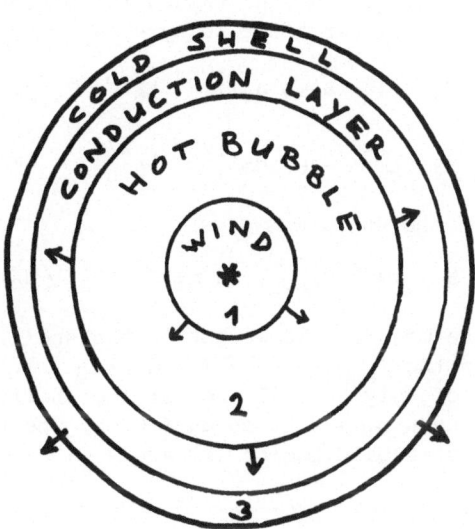

Figure 2. The structure of the bubble-complex during the snow-plough.

i: The energy of the wind is dumped into region (2) and used for heating and work done by the expansion. So

$$L_w = d\, E_2\, /\, dt + 4\, \pi\, R_2^2\, P_2\, (dR_2\, /\, dt) \tag{6}$$

where E_2 is the total heat energy of (2) since the kinetic energy of (2) can be neglected. We assume that region (2) is isobaric at a pressure P_2. The heat energy of the isobaric bubble is

$$E_2 = \int_{R_1}^{R_2} \frac{3}{2}\, n\, k\, T \cdot 4\, \pi\, r^2\, dr = \frac{3}{2} \int_{R_1}^{R_2} P_2\, 4\, \pi\, r^2\, dr \simeq 2\, \pi\, R_2^3\, P_2 \tag{7}$$

ii: The shell contains all the swept-up IS gas, so its mass is

$$M_3 = \frac{4}{3}\, \pi\, R_3^3\, \rho_o \simeq \frac{4}{3}\, \pi\, R_2^3\, \rho_o \tag{8}$$

where the last equality follows from the fact that the shell is thin, so $R_3 \simeq R_2$. The shell expands as a result of the gaspressure in the hot bubble, so

$$\frac{d}{dt}\left(M_3\, \frac{d\, R_3}{dt} \right) = 4\, \pi\, R_2^2\, P_2 \tag{9}$$

The solution of the equations (6) to (9) can be found by trying a solution of the kind $R_2 = \alpha t^\beta$. The results are expressed in terms of $L_{36} = L_w\, /\, 10^{36}$ erg/s and $t_6 = t\, /\, 10^6$ yrs. So the constants in the following equations give the values for a bubble complex from a wind of a typical O-star after 10^6 yrs. We find

$$R_2 \simeq R_3 = 28\ (L_{36}\, /\, n_o)^{1/5}\ t_6^{3/5} \qquad \text{pc} \tag{10}$$

$$v_2 \simeq v_3 = 17\ (L_{36}\, /\, n_o)^{1/5}\ t_6^{-2/5} \qquad \text{km/s} \tag{11}$$

The thickness of the shell is

$$d_3 = 3 \cdot 10^{-4}\ T_{shell}\ (\mu/m_H)^{-1}\ (L_{36}\, /\, n_o)^{-1/5}\ t_6^{7/5} \tag{12}$$

where μ/m_H is the mean atomic weight in the shell. If the shell is neutral, then the $T_{shell} = 10^2$ K and $\mu/m_H = 1$, so the shell has a thickness of only $3 \cdot 10^{-2}$ pc. If the shell is ionized then $T_{shell} = 10^4$ K and $\mu/m_H = 0.5$, so $d_3 \simeq 6$ pc. The temperature and density in the hot bubble region (2) are

$$T_2 = 1.6\ 10^6\ L_{36}^{8/35}\ n_o^{2/35}\ t_6^{-6/35} \qquad \text{K} \tag{13}$$

$$n_2 = 0.01\ L_{36}^{6/35}\ n_o^{19/35}\ t_6^{-22/35} \qquad \text{cm}^{-3} \tag{14}$$

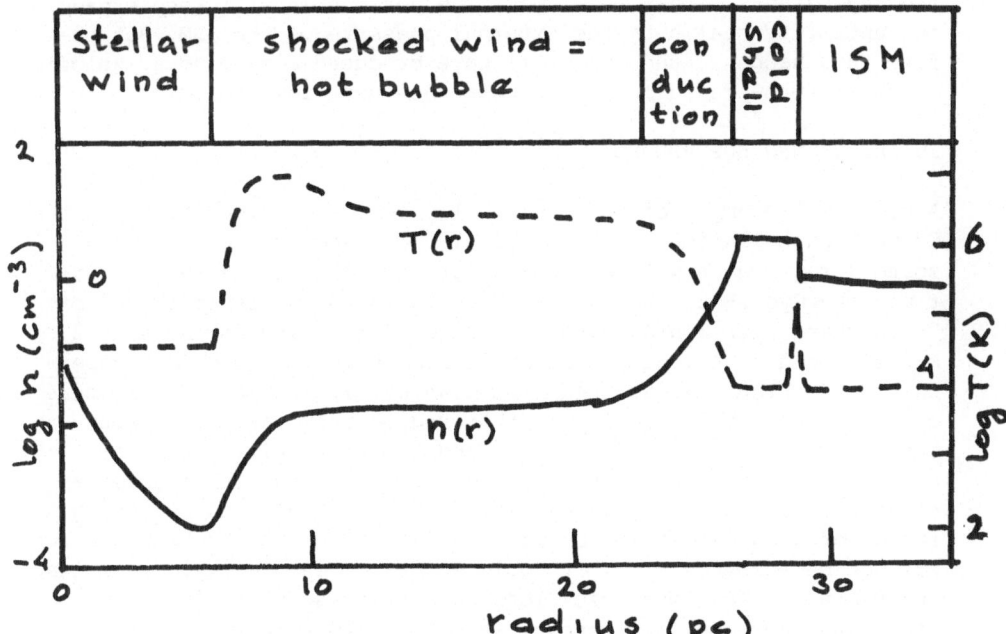

Figure 3. The temperature and density in a typical bubble of
$L_w = 1.27\ 10^{36}$ *erg/s,* $n_0 = 1\ cm^{-3}$ *after* $1\ 10^6$ *year (from Weaver et al., 1977).*

The hot bubble is a very weak soft X-ray source with a luminosity in the 44 Å - 70 Å band of

$$L_2 = 3 \cdot 10^{33}\ L_{36}^{37/35}\ n_o^{18/35}\ t_6^{16/35} \qquad erg/s \qquad (15)$$

The structure of a typical bubble after 10^6 yrs is shown in Figure 3 (from Weaver et al., 1977).

This model has a number of interesting observational consequences:

- The hot bubble at $T \sim 1.6\ 10^6$ K is a source of soft X-rays.
- The conduction region behind the cold shell may produce absorption lines which are typical for gas of $T \sim 10^5 - 10^6$ K, e.g. O VI and N V lines.
- When the cold shell has accumulated a sufficient amount of gas, it may trap the stellar UV photons. In this case there will be a partly neutral cold shell expanding in a fossile H II region. For a typical O7 star the ionization front will be trapped after $5 \cdot 10^6$ yrs.
- The high density shell may create H_2 sheets if the bubble expands into a cold molecular cloud.
- If the bubble grows inside an IS cloud and expands out of the cloud it will burst open and produce high velocity filaments.

d. Last phase: the dying star

After some 10^6 yrs the star will evolve into a red supergiant and L_w will drop drastically since the wind velocity of a red supergiant is only ∿ 20 km/s. The hot bubble will keep expanding untill it stalls by the external pressure of the ISM. But before this happens, the star will have produced a supernova explosion which will re-pressurize the bubble (See next lecture).

IV THE CYGNUS SUPERBUBBLE

A beautiful example of an interstellar bubble has been found in the Cygnus constellation. Cash et al. (1980) discovered an extended soft- X-ray region with a diameter of 13°. This X-ray region surround the Cyg OB 2 association which is known for its presence of 25 OB stars with a high mass loss rate. This association is at 2 kpc so the diameter of the X-ray region is 450 pc, indeed a <u>super-bubble</u>. In Figure 4 (adapted from Cash et al) I show the structure of this superbubble. I discuss this region from inside out with the interpretation of Cash et al. and Abbott et al., (1982).

    a. In the center is the Cyg OB 2 association with a diameter of 35 pc. This association is behind the Great Rift in Cygnus, a region of high extinction of about 5 magnitudes. If the association was not behind this Rift, it would be bright in the sky, resembling the Pleiades. The total mass loss rate of these stars is $2 \cdot 10^{-4}$ $M_\odot$/yr and the total wind energy is $L_w = 6 \cdot 10^{38}$ erg/s.

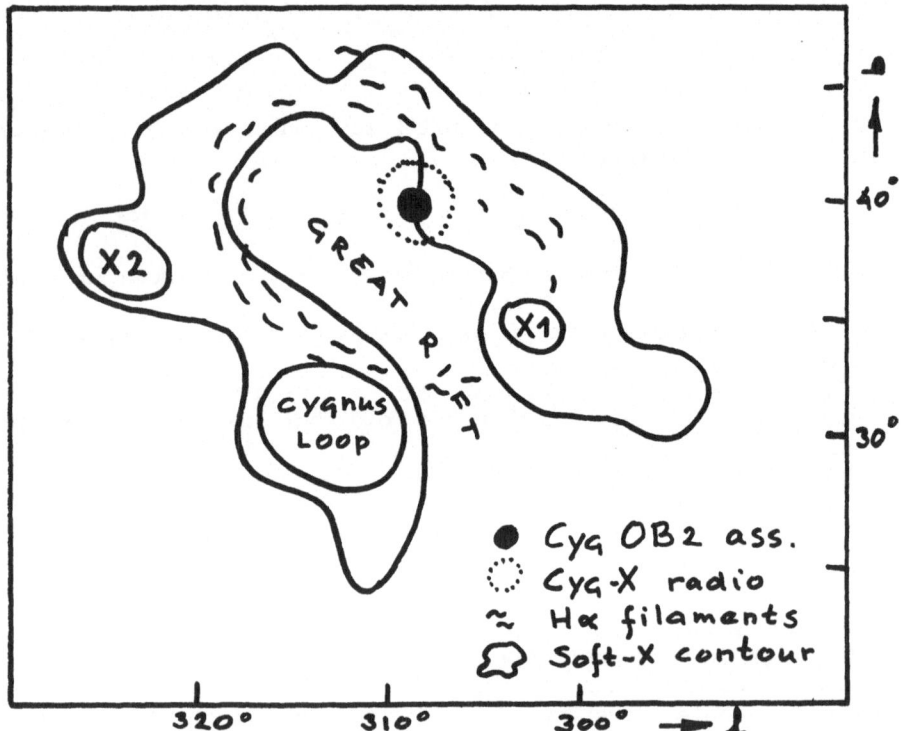

Figure 4. The structure of the Cygnus Superbubble (from Cash et al. 1980).

   b. Surrounding the associations is the Cyg X radio source
with a diameter of 100 pc. The radio spectrum is thermal, consisting of an unresolved background and 78 small sources. These components coincide with the regions of diffuse Hα and with Hα spots.
Abbott et al. proposed that this radiation comes from gas that was
evaporated from a nearby molecular cloud and ionized by the stellar
UV flux. It cannot be due to the hot bubble, since the bubble ex-
tends much further outwards.

   c. The soft-X region has a horse-shoe shape around the Great
Rift. Its diameter is about 450 pc. It is quite possible that the
X-ray region is in fact a filled sphere (in stead of a ring) with
the central part obscured by the Rift. The X-luminosity is $5 . 10^{36}$
erg/s, the emission measure is $1.5 \ 10^{59} \ cm^{-3}$, and the temperature
of the emitting gas is $2 . 10^{6}$ K. This is probably the hot bubble
(region 2 in Section III), powered by the collective winds of the
Cyg OB 2 stars. (The X-ray region also contains several discrete
sources, e.g. Cyg X1, Cyg X2 and the Cygnus Loop Supernova Remnant).

   d. A bright ring of Hα filaments with a diameter of about 500
pc surrounds the complex. This Hα emission could be due to the ionization of the shell of swept-up IS gas by the stellar UV photons.
One would expect the Hα filaments (i.e. the shell) to be outside
the X-ray region (the hot bubble), but the low spatial resolution
of the X-observations and the projection effects of a non-spherical
complex may confuse this.

   The predicted characteristics of a bubble, powered by the observed winds from the Cyg OB 2 association, and expanding into a
medium with a mean observed IS density of $n_O = 0.35 \ cm^{-3}$ are compared with the observations in Table 2 (from Abbott et al.). The
age of the bubble is assumed to be $2 . 10^6$ yrs, which is the evolutionary age of the association. The predictions agree remarkably
well with the observations, suggesting that the Cygnus Superbubble
is indeed a large scale example of wind bubbles. Other examples have
been studie by e.g. Bruhweiler and Gull (1980), Dopita and Wilson
(1981), Caulet et al (1982), De Boer and Nash (1982).

TABLE 2. PREDICTED AND OBSERVED PARAMETERS OF THE CYGNUS SUPERBUBBLE

	Predicted	Observed	
Radius	175	225	pc
Velocity	51	20-50	km/s
Luminosity	470	1300	$L_\odot$
Temperature	$5 . 10^6$	$2 . 10^6$	K
Particle density	0.01	0.02	cm^{-3}

REFERENCES

Abbott, D.C., Bieging, J.H., Churchwell, E., Cassinelli, J.P.:
    1980, Ap. J. 238, 196

Abbott, D.C., Bieging, J.H., Churchwell, E.: 1981, Ap. J. 250, 645

Bruhweiler, F.C., Gull, T.R.: 1980, in The Universe at Ultraviolet
    Wavelengths, ed. R.D. Chapman, NASA Conference Publication 2171

Cash, W., Charles, P., Bowyer, S., Walter, F., Garmire, G., Riegler,
    G.: 1980, Ap. J. 238, L71

Cassinelli, J.P.: 1979, Ann. Rev. Astr. Ap. 17, 275

Castor, J., McCray, R., Weaver, R.: 1975, Ap. J. 200, L107

Caulet, A., Deharveng, L., Georgelin, Y.M., Georgelin, Y.P.: 1982,
    Astron. Astrophys. 110, 185

Chiosi, C.: 1981, Astron. Astrophys. 93, 163

Conti, P.S.: 1978, Ann. Rev. Astron. Ap. 16, 371

De Boer, K.S., Nash, A.G.: 1982, Ap. J. 255, 447

De Loore, C., De Grève, J.P., Vanbeveren, D.: 1978, Astron. Astro-
    phys. Suppl. 34, 363

Dopita, M.A., Wilson, I.R.: 1981, in Effects of Mass Loss on Stellar
    Evolution, eds. C. Chiosi and R. Stalio, Reidel: Dordrecht, p.
    523

Garmany, C.D., Olson, G.L., Conti, P.S., Van Steenberg, M.E.: 1981,
    Ap. J. 250, 660

Lamers, H.J.G.L.M.: 1981, Ap. J. 245, 593

Lamers, H.J.G.L.M.: 1982, in Mass Loss from Astronomical Objects,
    ed. P.M. Gondhalekar (Rutherford Appleton Laboratory: Chilton)
    RL 82-075, p. 67

Morton, D.C.: 1968, Ap. J. 150, 535

Weaver, R., McCray, R., Castor, J., Shapiro, P., Moore, R.: 1977,
    Ap. J. 218, 377

Willis, A.J.: 1982, in Mass Loss from Astronomical Objects, ed.
    P.M. Gondhalekar (Rutherford Appleton Laboratory: Chilton)
    RL 82-075, p. 1

SUPERNOVAE AND SUPERNOVA REMNANTS

Henny J.G.L.M. Lamers

Astronomical Institute
Space Research Laboratory
Beneluxlaan 21, 3527 HS Utrecht
The Netherlands

I. INTRODUCTION

Supernovae are the result of explosions which terminate the life
of massive stars. The amount of energy released by the supernovae
in our galaxy is so large that these explosions provide the dominant energy for the heating of the hot (T $\gtrsim 10^6$ K) IS component
and for the kinetic energy of the large scale motions of the interstellar clouds. In addition, the ejected matter of supernovae is
the main source of the chemical pollution of the IS medium. In
this review I describe the different types of supernovae and their
frequency in our galaxy in Section II. The observations of supernova remnants are briefly described in Section III. The evolution
of the remnants is discussed in Section IV. In Section V I discuss
the effect of supernovae when they occur in stellar clusters. In
Section VI the overall effect of stellar winds and supernovae on
the interstellar medium is described.

In this review I will use the term Supernova (SN) to describe
the explosive phenomenon and the term Supernova Remnant (SNR) to
describe the interaction of the ejected matter with the ISM.

II. TYPES OF SUPERNOVAE AND THEIR FREQUENCY

There are at least two different types of SN which differ in the
shape of the light curve as well as in the spectral characteristics.
For reviews see: Zwicky (1969), Minkowski (1969), Clark (1982).

J. Audouze et al. (eds.), Diffuse Matter in Galaxies, 45–56.
Copyright © 1983 by D. Reidel Publishing Company.

a. Type I supernovae

This is a homogeneous group of SN. The typical lightcurve in the
visual is characterized by a phase of maximum luminosity lasting
about 30 days and an exponential decay when the brightness decreases by one magnitude per 80 days. The spectrum consists of
strong emission lines whose width corresponds to expansion velocities of 2 000 to 20 000 km/s. These SN are thought to result from
explosions of less massive population II stars as they are observed
in all kinds of galaxies, also in elliptical galaxies which only
consist of pop. II stars. The visual luminosity at maximum is about
-19 ± 0.7 magn.

b. Type II supernovae

This is a very inhomogeneous group. The lightcurve shows a shoul-
der which lasts about 100 days and then a rapid decline. The luminosity at maximum is about -17.7 ± 0.8 magn. The spectrum shows
strong Balmer emission lines with a width of about 6 000 km/s. The
type II SN are thought to result from population I stars; i.e.
young massive stars which are located mainly in the spiral arms of
galaxies. Typical lightcurves for type I and type II SN are shown
in Fig. 1. The mean energetic characteristics are summarized in
Table 1 (from Chevalier, 1977).

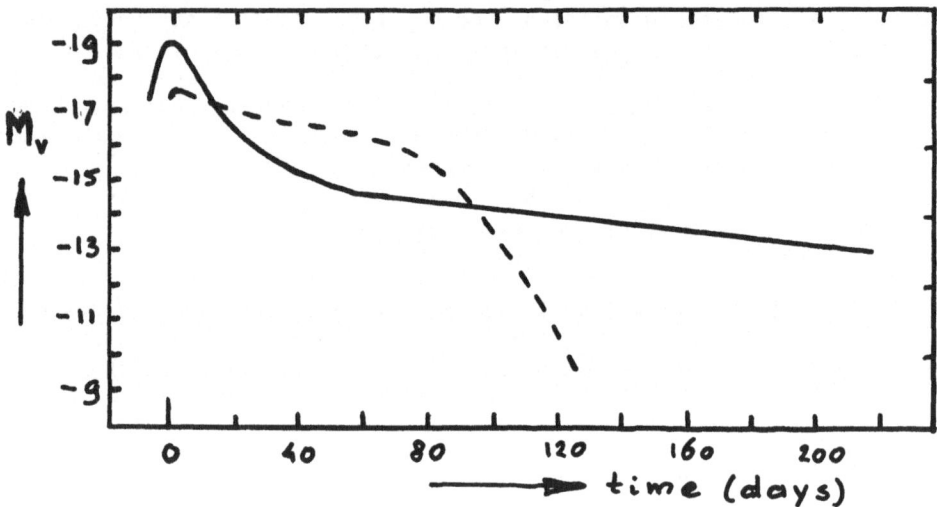

Figure 1: *Typical lightcurves of type I (full line) and type II
(dashed line) Supernovae (from Minkowski, 1969).*

Table 1.   ENERGETICS OF SUPERNOVAE

	Type I	Type II	
Ejected mass	0.5	5	$M_\odot$
Mean velocity	10 000	5 000	km/s
Kinetic energy	5.10^{50}	1.10^{51}	erg
Visual radiated energy	4.10^{49}	1.10^{49}	erg

c. The frequency of Supernovae in galaxies

The frequency of SN in external galaxies can be derived from statistics of SN-counts. This results in an estimated frequency of
one SN per galaxy per 20 to 50 years. In our own galaxy 8 SN have
been recorded in the literature over the last 2000 years. After
making large corrections for the estimated visibility of SN to
ancient observers and for the incompleteness of the ancient literature, Clark (1982) estimates that in our galaxy the supernova rate
must have been one in 20 years. This value is obviously very uncertain. The present best guess for the SN frequency is

$$1 \text{ SN per } 30 \, {}^{+20}_{-10} \text{ year per galaxy.} \tag{1}$$

It is interesting to calculate the total SN frequency in the Universe. Since there are about 10^{10} galaxies in the Universe we estimate a production rate of 10 SN per second!

    The estimated supernova rate in our galaxy can be compared
with the number of observed galactic supernova remnants. From the
number of observed SNR in our galaxy and the estimated observable
lifetime of 10^5 years per SNR, the production rate of the SNR in
our galaxy is found to be one in 80 years. This suggests that not
all SN explosions produce observable remnants. We return to this
problem in Section V.

III. OBSERVATIONS OF SUPERNOVA REMNANTS

Supernova remnants can be observed in the radio, optical and X-ray
wavelength regions (for reviews see: Poveda and Woltjer, 1969;
Woltjer, 1972). The young remnants of galactic SN are strong non-
thermal radio sources which usually show a spherical structure with
concentrations near the outer radius. A typical example is shown
in Figure 2. The young SNRs are also X-ray emitters. At optical
wavelengths the remnants show a strong filamentary structure near
the outer radius. The older SNR ($\sim 10^5$ yrs) are detected as extended non thermal radio sources only.

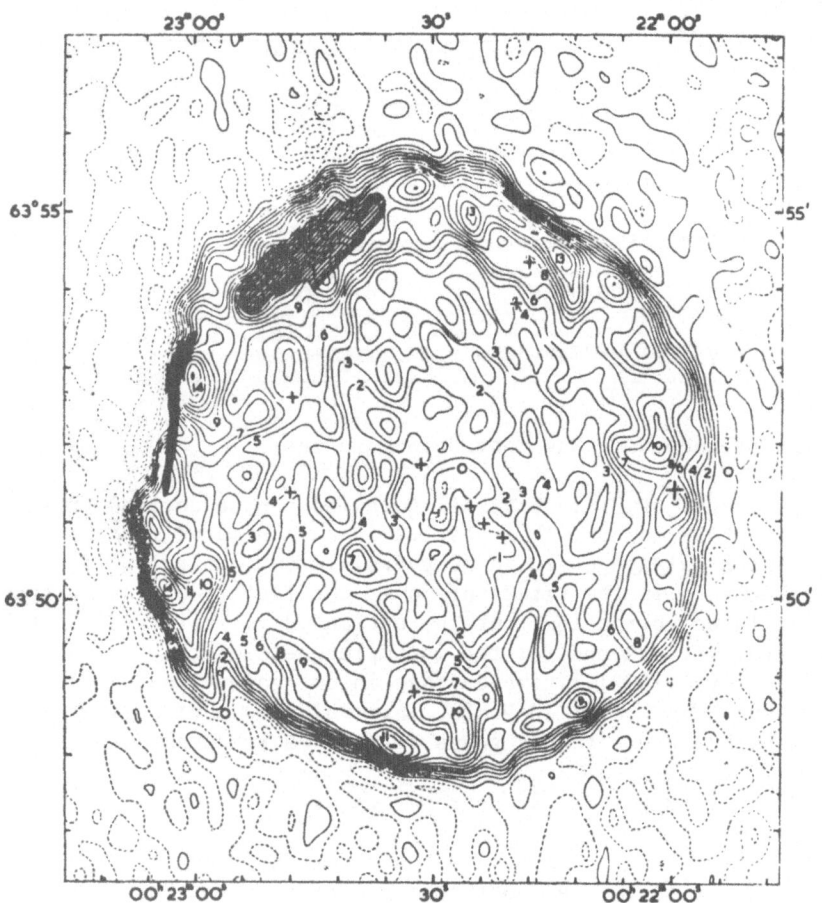

*Figure 2: The remnant of Tycho's SN observed at 1407 MHz. The op-
tical features are indicated by black and hatched areas
(from Minkowski, 1969).*

The expansion velocities of the remnants as derived from the
Doppler shifts of the optical filaments is of the order of 10^3 km/s
for the young remnants (Crab AD1054, Tycho AD1572, Keppler AD1604),
about 10^2 km/s for the Cygnus Loop ($\sim 3.10^4$ yrs) and probably goes
down to 10 km/s for the older remnants, but these are optically
too faint. These general properties are roughly in agreement with
the theoretical predictions of the evolution of a SNR.

IV. THE EVOLUTION OF SUPERNOVA REMNANTS

The ejection of a large amount of gas by a SN produces a bubble
in the ISM. This is quite similar to the formation of a bubble by

a stellar wind which was discussed in my previous lecture. The main difference between the two types of processes is the following: a stellar wind provides a constant energy and mass supply over a long period of 10^6 to 10^7 years, whereas a SN explosion is an instantaneous ejection when all the energy and mass is released at once. The total energy of the wind ($L_w.t$) however is comparable to the total kinetic energy (E_0) of a typical SN and is about 10^{51} ergs. For review articles see: Spitzer (1978, Chapter 12.2), Woltjer (1972) and Chevalier (1977).

In this section I will discuss the evolution of a typical SNR with an initial kinetic energy of $E_0 = 10^{51}$ ergs, an ejected mass of $M_0 = 0.25$ $M_\odot$ with an initial velocity of 20 000 km/s. The ISM is assumed to be homogeneous with a density of $n_0 = 1$ atom/cm^3 or $\rho_0 = \mu_0 m_H n_0 = 2.1 \ 10^{-24}$ g/cm^3.

a. Free expansion phase

At first the ejectum expands freely without any effect of the ISM. This phase ends when the amount of swept-up IS gas becomes comparable to M_0, which is when

$$4/3. \ \pi \ R^3 \ \rho_0 = M_0 \tag{2}$$

or R = 1.2 pc. This is already after 60 years.

b. Adiabatic expansion phase = Sedov phase

During this phase the mass of the expanding shell almost totally consists of swept-up IS material. As the shell expands supersonically into the ISM, the shock between the ISM and the shell will heat the shell to a temperature of 10^7 K. This temperature is so high that radiative cooling can be neglected. Therefore this is called the adiabatic phase. The structure of the SNR during the adiabatic phase is shown in Figure 3. It consists of an inner region which was heated when the shock passed through it and contains hot IS gas at low density. This hot bubble is surrounded by the expanding shell of swept-up IS gas. The shell expands because it still has a large fraction of the initial kinetic energy E_0. The equations for the structure and evolution of the SNR are derived: (i) The fraction K_1 of the original energy E_0 has been dissipated during the heating of the hot bubble. The rest $(1-K_1)E_0$ is the remaining kinetic energy of the expanding shell. The bubble has a mean gas pressure $\bar{P}_b$. The energy and pressure in the bubble are related by:

$$E_{th} = E_0 K_1 = \int_0^{R_b} \frac{3}{2} n_b kT_b 4\pi r^2 dr = \frac{3}{2} \bar{P}_b \frac{4}{3}\pi R_b^3 \tag{3}$$

So the mean pressure inside the bubble is

$$\bar{P}_b = E_0 K_1 / 2\pi R_b^3 \tag{4}$$

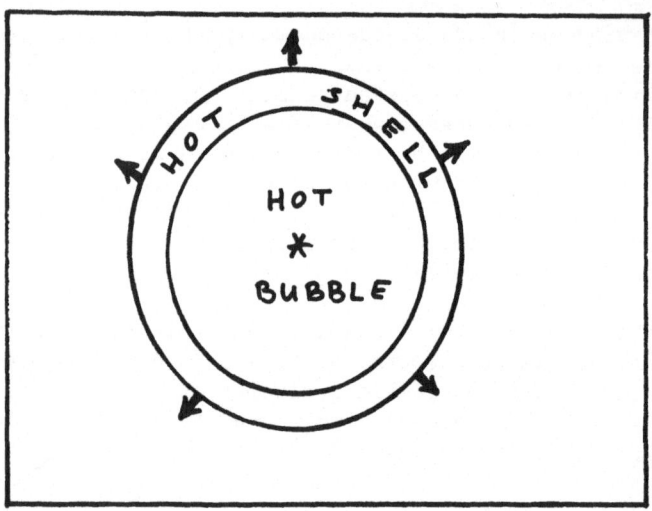

Figure 3: The structure of the SNR during the adiabatic phase

We want to express $\overline{P}_b$ in terms of the pressure P_s in the shell immediately behind the shock front.

(ii) The shell expands into the ISM by means of a strong adiabatic shock. The pressure P_s at the shock boundary is

$$P_s = 5/4 \; M^2 \; P_0 \qquad (5)$$

where $P_0 = n_0 \; k \; T_0$ is the pressure of the ISM and M is the Mach-number of the shock; which is related to the expansion velocity v_s of the shell:

$$M^2 = v_s^2 / (\gamma k T_0 / \mu m_H) \qquad (6)$$

The equations (5) and (6) can be combined to give an expression for the pressure P_s

$$P_s = 3/4 \; \rho_0 \; v_s^2 \qquad (7)$$

when we assume that $\gamma = 5/3$ for the ISM. The pressure in the bubble and behind the shock are related to one another by means of

$$P_s = K_2 \cdot \overline{P}_b \qquad (8)$$

(iii) Combining the equations (4), (7) and (8) provides an expression for the expansion velocity of the shell

$$v_s^2 = 2K_1 K_2 E_0 / 3\pi R_b^3 \rho_0 \stackrel{\sim}{=} E_0 / 3R_b^3 \rho_0 \qquad (9)$$

Modelcalculations show that $K_1 \sim 0.72$ and $K_2 \sim 2.13$. Since the shell is thin with $R_s \sim 1.1 R_b$ we can solve equation (9) by writing $v_s = dR_s/dt$. We find

$$R_s \sim (25E_o/12\rho_o)^{1/5} t^{2/5} = 0.32 \, n_o^{-1/5} \, t(\text{yrs})^{2/5} \, \text{pc} \tag{10}$$

and

$$v_s \sim (25E_o/12\rho_o)^{1/5} \cdot (2/5) t^{-3/5} = 1.28 \times 10^5 n_o^{-1/5} t(\text{yrs})^{-3/5} \, \text{km/s} \tag{11}$$

(iv) The temperature of the shock can be derived from the density at the shock $\rho_s = 4 \, \rho_o$ and the pressure P_s. This yields

$$T_s \sim \frac{3}{100} \, \mu (25E_o/12\rho_o)^{2/5} \, t^{-6/5} = 2.3 \, 10^{11} \, n_o^{-2/5} \, t(\text{yrs})^{-6/5} \, \text{K} \tag{12}$$

with $\mu = 0.63 \, m_H$. The radius, velocity, temperature and mass of this typical SNR as a function of time are listed in Table 2.

The adiabatic phase ends when the temperature has decreased to $T_s \lesssim 10^6$ K, because then radiative cooling becomes important. This is about after 3.10^4 years. The mass of the SNR is then about 1000 $M_\odot$, which is much larger than the originally ejected 0.25 $M_\odot$, so the remnant consists mainly of swept-up IS gas.

The observed Supernova Remnants are usually in this adiabatic phase.

c. Isothermal snow-plough phase

When the temperature of the shell has decreased to $T \lesssim 10^6$ K the radiative losses will rapidly cool the shell to a low temperature $(10^2 - 10^4$ K). The shock between the shell and the ISM will be isothermal which results in a high compression rate and produces a thin high density shell. The thin shell now keeps expanding be-

Table 2. THE EVOLUTION OF A TYPICAL ADIABATIC SNR

time (yrs)	R_s (pc)	v_s (km/s)	T_s (K)	M_s ($M_\odot$)
10^3	5.1	2030	$5.8 \, 10^7$	17
3.10^3	7.9	1050	$1.5 \, 10^7$	64
10^4	12.7	510	$3.6 \, 10^6$	270
3.10^4	19.8	260	$9.8 \, 10^5$	1000

cause of its momentum, so

$$M_s(t) \cdot v_s(t) \underset{\sim}{} \text{constant} \tag{13}$$

The constant is the momentum of the shell at the end of the adiabatic phase. The mass of the shell can be written as usual

$$M_s(t) = 4/3 \; \pi \; R_s(t)^3 \; \rho_o \tag{14}$$

and $v_s = dR_s/dt$. This yields a solution of the kind

$$
\begin{aligned}
R_s(t) \; (:) \; t^{1/4} \\
v_s(t) \; (:) \; t^{-3/4}
\end{aligned} \tag{15}
$$

Detailed models show that in reality $R_s(t) \; (:) \; t^{2/7}$ (Mc Kee and Ostriker, 1977; Shull and Silk, 1979; Chevalier, 1974). This snow-plough phase will end when the velocity of the shell becomes comparable to the mean velocity of the IS clouds, which is about 10 km/s. After that time the SNR breaks up by interaction with the ISM and looses its identity. When this happens the kinetic energy of the SNR is about $0.03 \; E_o$.

The different phases of the SNR are summarized in Table 3. Each SN adds about $0.03 \; E_o \underset{\sim}{} 3.10^{49}$ ergs of kinetic energy to the ISM. The other $0.97 \; E_o$ is used for heating of hot bubble, and part of it is later radiated away during the radiative cooling of the shell.

d. Instabilities in SNR

The phases discussed above were calculated for a smooth spherically symmetric SNR which expands into a homogeneous ISM. However, the

Table 3.   SUMMARY OF THE PHASES

Phase:	Free expansion	Adiabatic expansion	Isothermal expansion	
Until	$M_{swept-up} \underset{\sim}{} M_o$	$T_s \underset{\sim}{} 10^6$ K	$v_s \underset{\sim}{} 10$ km/s	
Duration	60	3.10^4	2.10^6	yrs
Radius	$R \sim t$	$R \sim t^{2/5}$	$R \sim t^{1/4}$	
Velocity	$v_s \sim v_o$	$v_s \sim t^{-3/5}$	$v_s \sim t^{-3/4}$	
Final radius	1	20	60	pc
Final velocity	20 000	260	10	km/s

observed SNR's always show a ring-like structure consisting of filaments with high density knots. See for example Fig. 2 or the figures in Minkowski (1969). These irregular structures can be a result of inhomogeneities in the originally ejected matter or in the
ISM (for a review: Chevalier, 1977).

The ejected matter can be clumpy. Such clumps have been observed in the young SNR Cas A. This remnant has high velocity
clumps which have an overabundance of argon, sulphur and oxygen,
which indicates that the clumps consist of originally ejected matter and not of swept-up IS gas. But even if the ejected material
is spherically symmetric, one can expect that the remnant will be
clumpy because it is subject to Rayleigh-Taylor instabilities,
especially in the young adiabatic phase. The older SNR's are more
sensitive to inhomogeneities in the ISM and may become clumpy when
they expand into a medium consisting of clouds and intercloud material.

V. SUPERNOVA REMNANTS FROM STELLAR CLUSTERS

In the preceding section we discussed the fate of a SNR in the
general ISM. In reality most of the SN explosions will occur in
highly disturbed regions of the ISM. Firstly, because the massive
star which produces the explosion at the end of its life has blown
a stellar wind bubble in an earlier phase of its life. So the SNR's
of massive stars will expand not into the general ISM, but inside
an IS bubble. Secondly, the massive stars live in clusters or associations, so a SN explosion will usually occur in a region where
the ISM is already disturbed by previous SNR from other stars. The
general effect of combined stellar wind bubbles and SN in clusters
has been studied by e.g. Bruhweiler et al. (1980) and Gull et al.
(1980).

Suppose a typical cluster with a few hundred stars in the
galactic plane has about 30 stars more massive than 15 $M_\odot$ and 180
stars with $8 < M < 15$ $M_\odot$. Bruhweiler et al. studied the simplified
history of such a cluster in three phases:
(i) the bubble phase. The 30 massive O-stars blow a large bubble
in the ISM by means of their combined stellar wind. This phase will
last about 3.10^6 years, which is the typical lifetime of an O-star.
At that time the bubble has an outer radius R $\sim$ 85 pc.
(ii) the first SN phase. After 3.10^6 years the 30 massive stars
explode inside the bubble. Since the bubble has a very low density
the SNR's can expand freely until they reach the outer edge of the
bubble. However, when the SNR's have expanded so much, they are
too diluted to produce a high temperature shock, so they will expand with a cold isothermal shock with R(t) (:) $t^{1/4}$. After 10^7
years the SNR's have increased the size of the bubble to R$\sim$170 pc.
(iii) the second SN phase. After 10^7 years the 180 stars with

$8 < M < 15$ $M_\odot$ will explode. The kinetic energy of these SN will provide an additional acceleration of the outer shell of the giant bubble. The shell continues to expand until its velocity has decreased to 10 km/s, when it will break-up and mix with the ISM. The maximum radius of the shell is about 170 to 700 pc, depending on the density of the ISM into which the shell expands.

In Table 4 the radius of the bubble at the end of the three phases is listed as a function of mean density of the ISM, which depends on the distance from the galactic center. The clusters are assumed to be in the galactic plane.

This model explains why the number of observed SNR's in the galaxy is smaller than the number expected on the basis of the SN frequency (see Section II). Since a SN will only produce an observable remnant if it expands into a medium with a density higher than about 0.1 atom/cm^3, the large fraction of the SN which explodes inside a superbubble will not produce an observable remnant. Bruhweiler et al. predict that only 20 to 30 percent of all the SN will produce remnants. This is in fair agreement with the small number of SNR's observed in the galaxy. This cluster model also predicts that about 30 percent of the volume of the galactic plane will consist of superbubbles.

Table 4.   THE RADIUS OF THE SUPERBUBBLES

R_{gal} (kpc)	n_H(ISM) (cm^{-3})	End of bubble phase (pc)	End of first SN-phase (pc)	End of second SN-phase (pc)
5	3	85	137	179
10	1	106	185	251
15	0.1	168	357	520

VI. THE OVERALL EFFECT OF STELLAR WINDS AND SN ON THE INTERSTELLAR MEDIUM

The main effect of SN and stellar winds on the ISM is the production of large bubbles which consist of a hot interior ($10^6 - 10^7$ K) at low density ($10^{-2} - 10^{-4}$ atoms/cm^3) surrounded by a thin dense shell of temperature $10^2 - 10^4$ K and density $10^1 - 10^3$ atoms/cm^3. The expanding shells are the main source of the kinetic energy of the ISM, e.g. the bulk motions of the IS clouds. The hot bubbles are the main source for producing the hot component of the ISM.

Abbott (1982) has estimated these effects on the basis of the number density of massive stars in the neighbourhood of the sun, and expressed the results in terms of energy input per kpc^2 in the galactic plane. He found that the SN and the stellar winds give an input of *kinetic* energy of 7.10^{37} ergs/s.kpc^2 and an input of energy for heating of the hot IS gas of 4.10^{38} ergs/s.kpc^2.

The cooling time of the hot gas is very long. For instance, at a mean IS gas pressure of 3.10^{-13} d/cm^2 (see Lequeux's lecture) the IS gas of 10^7 K has a density of 2.10^{-4} atoms/cm^3 and a cooling time of $1.7 \ 10^{10}$ years. In the course of the evolution of the galactic disc, the shells will be broken-up and mix with the ISM. The hot bubbles, however, will live almost forever and a large fraction of the IS space will consist of hot bubbles. The partly overlapping bubbles will create a system of IS tunnels of hot gas throughout the galactic plane (Figure 4). These effects are taken into account in the three-component model of the ISM by McKee and Ostriker (1977).

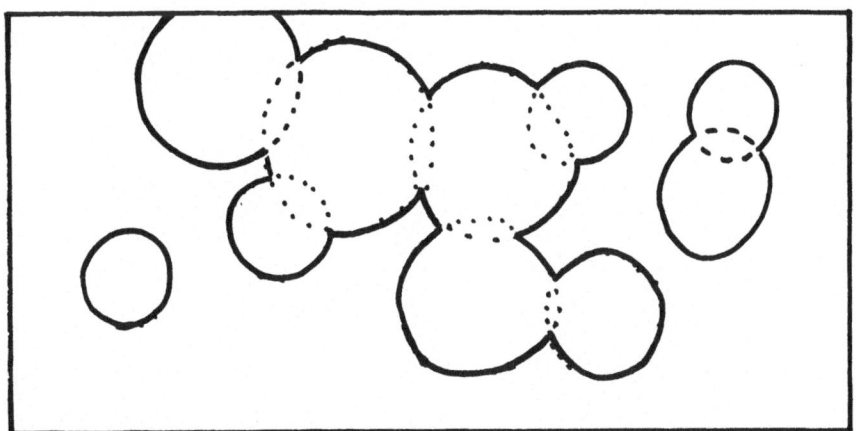

Figure 4: The system of partly overlapping superbubbles produces hot tunnels in the ISM.

REFERENCES

Abbott, D.C.: 1982, Astrophys. J. (preprint).
Bruhweiler, F., Gull, T., Kafatos, M., Sofia, S.: 1980, Astrophys. J. Letters, *238*, L27.
Chevalier, R.A.: 1974, Astrophys. J. *188*, 501.
Chevalier, R.A.: 1977, Ann. Rev. Astron. Astrophys. *15*, 175.
Clark, D.H.: 1982, in Mass Loss from Astronomical Objects, ed. P.M. Gondhalekar, Rutherford Appleton Lab., RL-82-075, p. 94.
Gull,T.R., Bruhweiler, F.C., Kafatos, M., Sofia, S.: 1980, in The

Universe at Ultraviolet Wavelengths, ed. R.D. Chapman, NASA
    Conf. Publ. 2171, p. 679.
McKee, C.F., Ostriker, J.P.: 1977, Astrophys. J. *218*, 148.
Minkowski, R.: 1969, in Supernovae and their Remnants, eds. P.J.
    Brancazio, A.G.W. Cameron (Gordon and Breach Sc. Publ., New
    York) p. 23.
Poveda, A., Woltjer, L.: 1969, in Supernovae and their Remnants,
    eds. P.J. Brancazio, A.G.W. Cameron (Gordon and Breach Sc.
    Publ., New York) p. 43.
Shull, J.M., Silk, J.: 1979, Astrophys. J.
Spitzer, L.: 1978, Physical Processes in the Interstellar Medium,
    (J. Wiley and Sons, New York) p. 255.
Woltjer, L.: 1972, Ann. Rev. Astron. Astrophys. *10*, 129.
Zwicky, F.: 1969, in Supernovae and their Remnants, eds. P.J. Bran-
    cazio, A.G.W. Cameron (Gordon and Breach Sc. Publ, New York)
    p. 1.

INTERSTELLAR HELIUM AND DEUTERIUM

A. VIDAL-MADJAR
L.P.S.P.   -   B.P.   10   -   91370   Verrières-le-Buisson,
France.

ABSTRACT

   The importance of helium and deuterium is underlined: both are probably ashes of the primordial nucleosynthesis and represent some of the most abundant elements in the universe. For both elements the different observational methods used in the interstellar medium are mentioned and the most precise ones described in more details. A discussion of the results in comparison with current theories shows that many more observations are needed.   Typical future possibilities are suggested.

J. Audouze et al. (eds.), Diffuse Matter in Galaxies, 57–94.
Copyright © 1983 by D. Reidel Publishing Company.

I - <u>INTRODUCTION</u>

The abundance of the elements in the diffuse interstellar medium is now relatively well known.   Among them, helium and deuterium are important for similar reasons.

Both are among the most abundant ones, helium representing about 25% of the mass while deuterium is roughly as abundant as the well known abundant elements, i.e. C.N.O.

Both also were very probably formed during the first few minutes of the universe and thus represent cosmological probes.

Both should represent good tracers of galactic evolution since one is formed inside stars through hydrogen burning while the other is almost always destroyed in stars.

Finally both are observed with the highest degree of confidence in the diffuse interstellar medium.

Their formation in the frame of the standard Big Bang model is presented in section II along with a discussion of their possible evolution.   In sections III and IV, the helium and deuterium observations are presented while conclusions and possible future observations are suggested in section V.

II - FORMATION AND EVOLUTION

The main site of formation of helium and deuterium is the early universe according to the so-called Big Bang theory as originally described by Alpher, Bethe and Gamov (1948). One of their assumptions was that the matter was supposed initially to be essentially neutrons. From that condition it is easy to produce helium but, as underlined by Sciama (1971), this leads to an unacceptable cosmological black body temperature of 25°K today. Hayashi (1950) showed that the neutron abundance is not a free initial parameter but is determined by thermal equilibrium at $T > 10^{10}$ K.

In effect, following Peebles (1971), at $T > 10^{10}$ K the radiation and the electrons have to be in thermal equilibrium because the thermal energy is high enough to permit electron pair formation through reactions like

$$\gamma + \gamma \rightarrow e^+ + e^-$$

and because the rate of such a reaction is much shorter than the expansion rate of the universe.

Under these conditions, the neutron-proton abundance ratio is fixed by the reactions

$$n \rightleftarrows p + e^- + \bar{\nu}$$

$$e^+ + n \rightleftarrows p + \bar{\nu}$$

$$\nu + n \rightleftarrows p + e^-$$

and at equilibrium is simply given by

$$\frac{n}{p} = \exp \left[- \frac{(m_n - m_p) \, c^2}{kT} \right] \tag{1}$$

where m_n and m_p are the neutron and proton masses. The energy $(m_n - m_p) \, c^2$ is of the order of 1.3 MeV, a value equivalent to the thermal energy at $T \sim 10^{10}$ K. Above 10^{10} K the neutron-proton abundance ratio is fixed by (1) while below 10^{10} K the electron-positron pairs disappear quickly and dump their energy into the radiation. The neutron-proton equilibrium is then broken and the neutron-proton abundance ratio is "frozen in" at $T = 10^{10}$ K, i.e.

$$\frac{n}{p} \sim e^{-1.5} \sim 0.2.$$

Knowing then the neutrino energy distribution (decoupled from matter at $T \sim 10^{11}$ K) and taking into account the free neutron decay reaction

$$n \rightarrow p + e^- + \bar{\nu}$$

whose half life is of the order of 11 mn, it is possible to calculate the evolution of the neutron abundance ratio. The sequence of events is given in Table 1 (from Peebles, 1971).

Table 1

Early universe evolution

time s	radiation temperature 10^{10} K	neutrino temperature 10^{10} K	$\frac{n}{(n+p)}$	event	
0.0001	100	100	0.496		
0.0109	10	10	0.462	neutrino decouple	thermal equilibrium
0.273	2	1.996	0.330		
1.102	1	0.992	0.238	matter decouple	
182.	0.1	0.074	0.130		neutron decay
296.	0.08	0.058	0.116	deuterium formation	
535.	0.06	0.043	0.089	helium formation	

As shown earlier, the neutron abundance ratio is "frozen in" at 10^{10} K until a few hundred seconds have passed when neutron decay starts to be appreciable. The build-up of heavier particles through nuclear reactions will thus take place in a mixture presenting a neutron abundance ratio in the range 0.1-0.2.

The key nuclear reaction is then

$$n + p \rightleftarrows d + \gamma$$

leading to the formation of deuterium. Once it is formed, it burns rather quickly into helium and almost all the neutrons are then locked into helium nuclei.

But the formation of deuterium is possible only under two conditions : i) that the temperature is less than 10^9 K because above that temperature there are too many photons able to des-integrate deuterons and ii) that the density is high enough to give a production rate faster than the expansion rate of the universe.

The equilibrium abundance ratio of deuterium is defined by the following relation

$$\left(\frac{x_n \, x_p}{x_d}\right)_{equil} = \frac{4}{3} \frac{(2\pi \, kT)^{3/2}}{(2\pi \, \hbar)^3 \, n} \left(\frac{m_n \, m_p}{m_d}\right)^{3/2} e^{-B/kT} \qquad (2)$$

where B is the deuterium binding energy of 2.225 MeV, $x_i = N_i/N$ is the abundance ratio of any type of particle versus all nucleus N (one should have $x_n + x_p + 2\, x_d = 1$), and n is the nucleon number density : $n = N/V$. This nucleon number density is related to the present nucleon density n_0 in the universe by the relation

$$n = \left(\frac{T}{2.7}\right)^3 n_0 \qquad (3)$$

Over a reasonable range of nucleon density it could be shown that the equilibrium ratio of deuterium (relation (2)) is near unity for $T \sim 0.08\ 10^{10}$ K. At this temperature, as shown in Table 1, the neutron abundance ratio is

$$\frac{n}{n+p} \sim 0.12$$

which gives a helium abundance by mass of

$$Y = \frac{2\ n}{n+p} \sim 0.24$$

since once deuterium is formed, it burns extremely quickly to form helium where almost all neutrons are finally locked.

It is remarkable to note that nuclear burning (to form deuterium) can start (around $T \sim 0.08\ 10^{10}$ K) just when the age of the universe is comparable to the neutron half life, i.e. when neutrons are starting to decay freely.

Because helium is formed from deuterium that can appear only in a small range of temperatures, the helium abundance ratio is quite insensitive to the model parameters and particularly to the present nucleon density.

To evaluate the deuterium abundance much more elaborate calculations must be carried out. Wagoner (1973) included a large number of nuclear reactions in his calculations involving all elements up to oxygen. His calculations illustrate (see Figure 1) again the early universe evolution as described in table 1 for a "standard" Big Bang model, i.e. with the following assumptions:

1)   the universe is homogeneous and isotropic;
2)   general relativity is assumed;

3)   only known particles are included;
4)   the baryon number is positive;
5)   the lepton numbers are less than the number of photons.

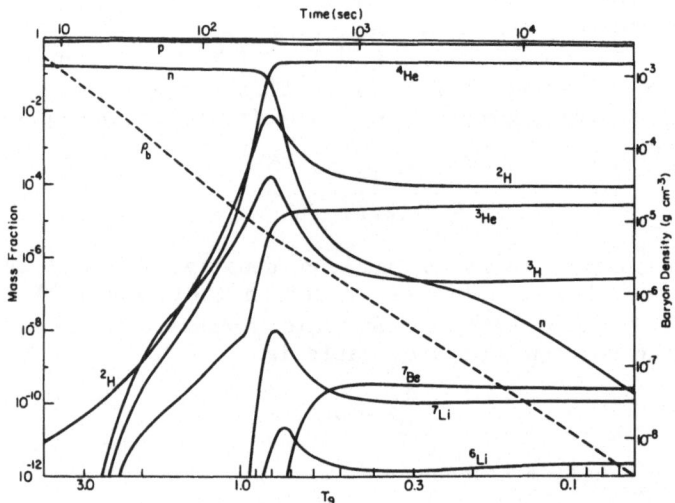

Fig. 1.   The evolution of nuclear mass fractions X(i)
and baryon density ρ_B (dashed curve) during the expansion
of the standard Big Bang model (from Wagoner, 1973).

From this calculations it is possible to evaluate the abundance
of the different synthesized elements as a function of the
present baryon density of the universe as shown on Figure 2 also
from Wagoner (1973).   This Figure illustrates also what was
previously described: once deuterium is formed, it burns almost
immediately into helium in which all neutrons are finally
locked, inducing a fairly constant helium abundance ratio.   At
lower densities the deuterium can survive slightly better until
the universe expansion protects it from burning; this induces
larger deuterium abundances.   At densities even lower than the
one presented on Figure 2 the deuterium formation could not be
started and thus deuterium, helium and all other species
abundances should be smaller.

This type of calculation shows that the helium abundance
represents a strong constraint on the model itself while the
deuterium abundance is probably the best tracer of the present
baryon density of the universe.

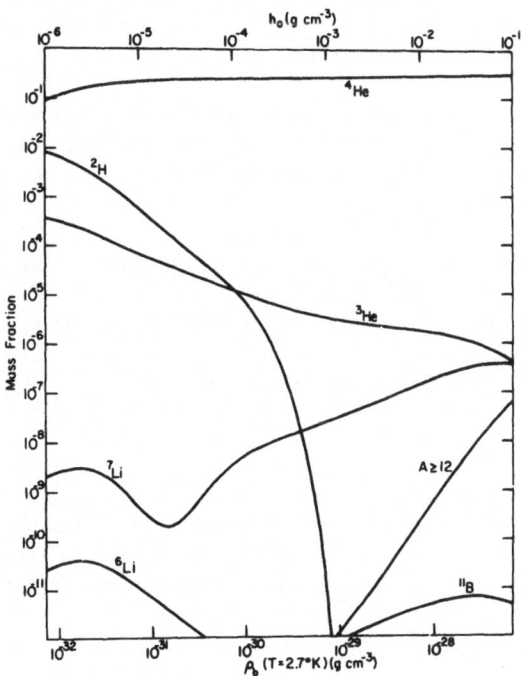

Fig. 2.   Final abundances produced in the standard
Big Bang model (from Wagoner, 1973).

     More recent calculations by Beaudet and Yahil (1977)
confirm these results and deviations from the standard model as
shown by Schramm and Wagoner (1977) seem to be hardly acceptable if one wants to reproduce the helium abundance ratio.
However, David and Reeves (1980) have shown that with arbitrary
lepton numbers almost any abundance ratio of the light elements
could be predicted, but Schramm and Steigman (1979) underline
that the required lepton numbers should be so large that they
seem highly improbable.

     Another possibility is that the number of neutrinos could
be larger than the two types introduced in the standard Big Bang
calculations. This effect reviewed in detail by Steigman (1979)
and Taylor (1980) should affect only the helium abundance ratio
and not the deuterium one. Steigman, Schramm and Gunn (1977)
and more recently Yang et al. (1979) have used the observed
helium abundance to deduce on the contrary that only one
additional stable neutrino could be accepted in the frame of the
Big Bang model.

In any case it seems clear that the situation in the early universe is favorable for the light element production; it is important now to try to see if other formation sites are possible and thus what evolution we may expect.

Reviews of other sources of the light elements have been made by Reeves et al. (1973), Reeves (1974), Laurent (1978), Reeves and Meyer (1978), and Austin (1980), and we shall mention here these possibilities:

1)   Shock waves in Type II supernovae may form deuterium as suggested by Hoyle and Fowler (1973) and Colgate (1973), but the energy requirements seem high (Reeves, 1974).

2)   Spallation due to galactic cosmic rays; but as shown by Meneguzzi et al. (1971) too much Li, Be and B would be produced.

3)   Deuterium production from neutrons ejected from neutron star near black holes as suggested by Lattimer and Schramm (1974). But as shown by Epstein et al. (1976) most of the produced deuterium should be immediately burned.

4)   Pre-galactic formation in hypothetical supermassive objects proposed by Truran and Cameron (1971). But Wagoner (1969, 1973) showed that if deuterium were produced with the observed abundance this would produce too much helium.

5)   Spallation due to pre-galactic cosmic rays due to quasar-like events before heavy elements are formed is proposed by Epstein (1977). Although this avoids the Li-Be-B over production as mentioned in (1), it places some constraints on the diffuse γ-ray background. To be compatible with the γ-ray background observations the formation sites should be either heavily shielded or at very early epochs ($Z \sim 400$).

6)   In stellar flares deuterium production is possible since it is observed in solar flare (Chupp et al., 1973; Anglin et al., 1973; Anglin, 1975). As underlined by Coleman and Worden (1976) production from flares in M type dwarf stars could be important due to their large number. But again a too large production of other elements like ^7Li contradicts this mechanism.

7)   Finally inside main sequence stars it is clear that deuterium is burned since no deuterium is seen in the sun (Beckers, 1975): as shown by Geiss and Reeves (1972) it is probably transformed into ^3He. Also it is well known that helium is formed inside main sequence stars.

Except for the last one, all these mechanisms look like complications which do not seem to be necessary although some of them should be kept in mind if observations seem to ask for them.

Thus the evolution of the deuterium and helium abundance could be tentatively described by ignoring all these points, but formation during the early universe followed by destruction of deuterium and build up of helium when matter passes into stars during galactic evolution. Calculations of galactic evolution were performed by Audouze and Tinsley (1974) showing that in effect one may predict the evolution of these species but with assumptions which are still far from being precisely defined by observations, like the initial mass function of stars, mass loss by massive stars, the instant recycling hypothesis or the composition of the infalling gas on a galaxy. This subject was reviewed by Audouze and Tinsley (1976) and re-analyzed by Vigroux et al. (1976), Chiosi et al. (1978), Lequeux et al. (1979), Vigroux (1979), Tinsley (1980), and Audouze and Vauclair (1980). They show that the reduction of deuterium abundance is a rather small effect (within a factor of two) during the life of our galaxy, while the models of chemical evolution of galaxies can predict that the helium to heavy elements enrichment ratio $\Delta Y/\Delta Z$ (where Y is helium abundance by mass and Z is heavy element abundance by mass) should be in the 1 to 4 range.

All these theoretical predictions should be compared now with observations.

III - <u>OBSERVATION OF INTERSTELLAR HELIUM</u>

Although it is well known that helium is observed in stars (and even discovered in the sun), as argued by Austin (1980) or Pagel (1982), these estimations of helium abundances involve detailed and complex stellar models for evolved stars such that one could question the precision of these evaluations, particularly because some observations fall well outside the expected range.

Fortunately, the evaluation of helium abundance in the interstellar medium seems to be more reliable. In effect, Pagel (1982) quotes precision better than 10 per cent in the helium abundance evaluation toward some H II regions.

Helium can be observed in H II regions through re-combination lines either in the optical or in the radio part of the spectrum. Because in the radio region only He^+ can be observed (see e.g. Churchwell et al., 1978) additional assumptions about the ionisation in H II regions are needed in order to deduce the helium abundance. For that reason we will limit here our description to the most reliable results obtained from optical observations of He^+ and He^{++} in H II regions inside our galaxy as well as in other galaxies.

As described e.g. by Seaton (1960), the emission lines and continuum from a low density gas exposed to dilute ultraviolet radiation are due to physical processes which, once analyzed, should lead us to an evaluation of the gas temperature, density and composition.

These physical processes are essentially   i) the atomic photo-ionization by absorption of UV photons

$$H + h\nu \rightarrow H^+ + e^-$$

and also

$$He + h\nu \rightarrow He^+ + e^-$$

which represents the source of random kinetic energy in the nebula because elastic collisions between charged particles is very effective in maintaining a Maxwell velocity distribution, i.e. in the nebula a kinetic temperature exists;   ii) loss of kinetic energy comes from inelastic collisions followed by emission of radiation.   In some of these collisions low lying metastable states of ions and atoms are populated permitting the emission of forbidden lines.   Because the probability of collisional deactivation is not negligible,   forbidden lines of a given ion have intensities which are function of both Te and Ne, the electronic temperature and number density.   Their study should thus give an evaluation of these two parameters.

Also should appear in emission permitted lines simply due to recombination processes, e.g.

$$H^+ + e^- \rightarrow H_{(nl)} + h\nu$$

and the same for He^{++} and He^+.

These lines are strong only if the abundance of the considered element is large, i.e. recombination lines of H^+, and He^+ and He^{++} could be expected.

The line intensities have been calculated by several authors using simplifying assumptions until the most complete ones made by Brocklehurst (1971) that we will briefly describe here. His formulation is the following: he equates processes per cm^{-1} and per s^{-1} which populate and depopulate a quantum level nl of a given specie

$$N_e N_+ \alpha_{nl} + \sum_{n'=n+1}^{\infty} \sum_{1'=1\pm1} A_{n'1',nl} N_{n'1'}$$

$$+ \sum_{1'=1\pm1} N_e N_{nl'} C_{nl',nl}$$

$$+ \sum_{n'=n_0}^{\infty} \sum_{1'=1\pm1} N_e N_{n'1'} C_{n'1',nl}$$

$$= N_{nl} [A_{nl} + \sum_{1'=1\pm1} N_e C_{nl,nl'}$$

$$+ \sum_{n'=n_0}^{\infty} \sum_{1'=1\pm1} N_e C_{nl,n'1'}] \qquad (4)$$

where the radiative recombination coefficient of level nl is α_{nl} and the radiative transition probabilities from level n'1' to level nl is $A_{n'1',nl}$. A_{nl} is the total radiation transition probability from level nl. $C_{nl',nl}$ is the collision rate for redistribution of angular momentum and $C_{n'1',nl}$ the one for redistribution of energy. The coefficients, α nl, $C_{nl',nl}$, $C_{n'1',nl}$, are functions of the electronic temperature Te through the Maxwellian velocity distribution.

Te and Ne being given and all the other coefficients being calculated, the population of the nl level n_{nl} can be evaluated. The tabulated values of deduced line intensities should then be accurate to the 5 percent level permitting precise abundance evaluations.

Observations of H and He recombination lines yielding to precise helium abundance evaluations were made by several authors (see e.g. Figure 3 from Rayo et al., 1982), but to compare the observations to theoretical predictions several problems arise that we should describe here.

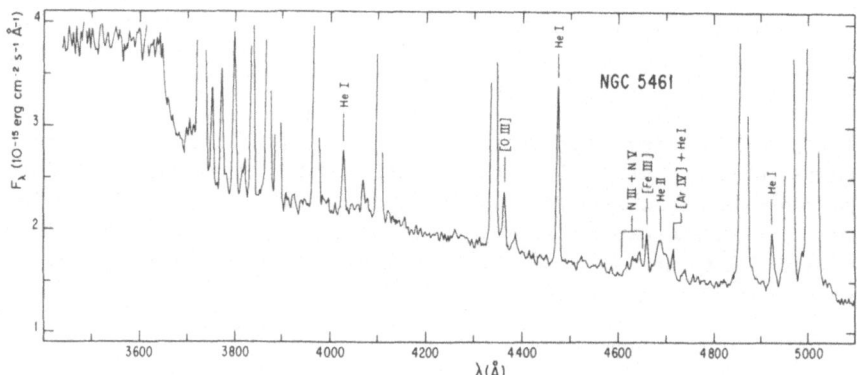

Fig. 3.   An example of an emission line spectrum of a H II region in the M 101 galaxy (Rayo et al., 1982) where some of the faintest emission lines are labelled. Note the Balmer discontinuity near 3600 Å.

As discussed by Peimbert and Torres-Peimbert (1976) the first effect to take into account is interstellar reddening. This affects the observed line intensities and the correction is made according to

$$\log [I(\lambda)/I(H\beta)] = \log [F(\lambda)/F(H\beta)] + C(H\beta) \; f(\lambda) \qquad (5)$$

where $F(\lambda)$ is the observed line flux at the wavelength λ, corrected for atmospheric extinction, $C(H\beta)$ is the logarithmic reddening correction at $H\beta$ and $f(\lambda)$ is the reddening function, normalized at $H\beta$ and derived from a normal extinction law (Whitford, 1958).   Knowing the $f(\lambda)$ function, the $C(H\beta)$ coefficient is adjusted by comparing the observed Balmer decrement to the one computed by Brocklehurst (1971). Further adjustment could be made by comparing also the observed $F(H\alpha)/F(H\beta)$ ratio to the predicted one.

Table 2 taken from Lequeux et al. (1979) shows the identification of the optical lines observed along with the reddening correction and the observed line intensities in log $[I(\lambda)/I(H\beta)]$.

Note here that the "standard" reddening function and the calibration necessary to connect two parts of the spectrum taken independently (a necessity of the Image Dissector Scanner System) represent the major cause of systematic errors. They are difficult to evaluate and overcome and are probably related to the observer (see a discussion in Lequeux et al., 1979).

Table 2

Absolute line intensities and reddening corrections; the line intensities are given in log $I(\lambda)/I(H\beta)$, $f(\lambda)$ is the reddening function, $C(H\beta)$ is the logarithmic reddening correction, and $I(H\beta)$ is the logarithm of the intrinsic flux in erg cm^{-2} s^{-1}.

	Ident.	$f(\lambda)$	NGC 4449 39	NGC 6822 V	NGC 6822 X	IC 10 1	IC 10 2	II Zw 70	II Zw 40	I Zw 18
3727	[O II]	+0.315	+ 0.48	+ 0.16	+ 0.30	+ 0.17	+ 0.16	+ 0.49	+ 0.05	− 0.42
3798	H 10	+0.290	· ...	− 1.30	...	...	...	...	...	...
3835	H 9	+0.280	− 1.14	− 1.12	− 1.18	...	...	...	...	...
3869	[Ne III]	+0.270	− 0.61	− 0.38	− 0.44	− 0.63	− 0.44	− 0.41	− 0.20	− 0.71
3889	He I + H 8	+0.265	− 0.70	− 0.68	− 0.65	− 0.76	− 0.76	− 0.85[a]	− 0.70	− 0.75
3967 + 3970	[Ne III] + H 7	+0.235	− 0.65	− 0.56	− 0.57	− 0.68	− 0.58	− 0.68[a]	− 0.46	− 0.68
4026	He I	+0.225	...	− 1.71	...	...	...	...	...	...
4068 + 4076	[S II]	+0.210	− 1.80	− 1.98:	...	...	...	...	...	...
4102	Hδ	+0.200	− 0.59	− 0.60	− 0.59	− 0.56	− 0.62	− 0.64[a]	− 0.61	− 0.58
4340	Hγ	+0.135	− 0.33	− 0.32	− 0.31	− 0.34	− 0.33	− 0.33	− 0.27	− 0.32
4363	[O III]	+0.130	− 1.77	− 1.28	− 1.39	− 1.40:	− 1.36:	− 1.25	− 0.93	− 1.13
4472	He I	+0.105	− 1.46	− 1.43	− 1.37	...	− 1.40:	− 1.50	− 1.38:	− 1.38:
4658	[Fe III]	+0.050	− 2.10	− 2.00:	...	...	...	− 1.75	...	...
4686	He II	+0.045	− 1.98 <	− 2.26 ≲	− 2.00 <	− 1.80 <	− 1.97	− 1.62 <	− 1.68 <	− 1.4
4713	[Ar IV] + He I	+0.035	...	− 2.02:	...	...	...	...	...	...
4861	Hβ	+0.000	+ 0.00	+ 0.00	+ 0.00	+ 0.00	+ 0.00	+ 0.00	+ 0.00	+ 0.00
4921	He I	−0.015	− 2.15	− 1.95	...	...	...	...	...	...
4959	[O III]	−0.020	+ 0.06	+ 0.26	+ 0.20	+ 0.10	+ 0.25	+ 0.12	+ 0.39	− 0.16
5007	[O III]	−0.030	+ 0.56	+ 0.73	+ 0.69	+ 0.58	+ 0.73	+ 0.62	+ 0.88	+ 0.33
5198 + 5200	[N I]	−0.075	− 2.40:	...	...	...	...	...	...	...
5876	He I	−0.210	− 1.02	− 0.97	− 0.95	− 1.00	− 1.02	− 1.01	− 1.09	− 1.06
6300	[O I]	−0.285	− 1.61	− 1.97	...	...	...	− 1.48	− 1.62	...
6311	[S III]	−0.290	− 2.03	− 1.86	...	...	− 1.65	− 1.69	...	...
6363	[O I]	−0.300	− 2.10	...	...	...	...	...	− 2.15	...
6563	Hα	−0.335	+ 0.45	+ 0.45	+ 0.45	+ 0.45	+ 0.45	+ 0.45	+ 0.45	+ 0.45
6584	[N II]	−0.340	− 0.83	− 1.21	− 1.23	− 0.97	− 0.94	− 0.88	− 1.14 <	− 1.3
6678	He I	−0.360	− 1.62	− 1.60	− 1.65:	− 1.60	− 1.57	− 1.66:	− 1.58	...
6717	[S II]	−0.370	− 0.84	− 1.20	− 1.11	− 0.99	− 1.05	− 0.71	− 1.25	− 0.97
6731	[S II]	−0.370	− 0.98	− 1.35	− 1.37	− 1.14	− 1.19	− 0.81	− 1.33	...
7065	He I	−0.400	− 1.83	− 1.68	...	...	− 1.46	− 1.74	− 1.43:	...
7136	[Ar III]	−0.410	− 1.17	− 1.06	− 1.02	− 1.01	− 0.93	− 1.22	− 1.19	...
7320 + 7330	[O II]	−0.435	− 1.49	− 1.66	...	− 1.49::	...	− 1.33	...	...
$C(H\beta)$			0.6	0.8	0.8	1.2	1.5	0.5	1.1	0.1
$I(H\beta)$			−12.35	−12.05	12.56	−12.47	−12.16	−12.87	−12.29	−13.68

[a] Underlying absorption present

(from Lequeux et al., 1979).

After having evaluated the electronic temperature Te from [0 III] lines and the electron density from the [S II] lines it is possible to evaluate the He^+ and He^{++} abundance by comparison with Brocklehurst (1971, 1972) calculations.

Then to derive the helium abundance ratios, several corrections should be made; one has i) to take into account possible temperature fluctuations along the line of sight which in any case affect only slightly the helium abundance evaluation (see Osterbrock, 1974; Peimbert and Torres-Peimbert, 1977); ii) to evaluate the unobserved He^0 abundance ratio since the helium abundance ratio is given by

$$\frac{N(He)}{N(H)} = \frac{N(He^0 + He^+ + He^{++})}{N(H^+)} \; ;$$

for that purpose, ionization correction factor could be evaluated from the observed $N(0^+)/N(0)$ ratio (Peimbert et al., 1974) or by extrapolating the relation between $N(He^+)/N(H^+)$ and $N(0^{++})/N(0^+ + 0^{++})$ toward higher ionisation states, i.e. toward $N(0^{++})/N(0^+ + 0^{++}) = 1$ (see Figure 4, taken from Peimbert and Torres-Peimbert, 1976).

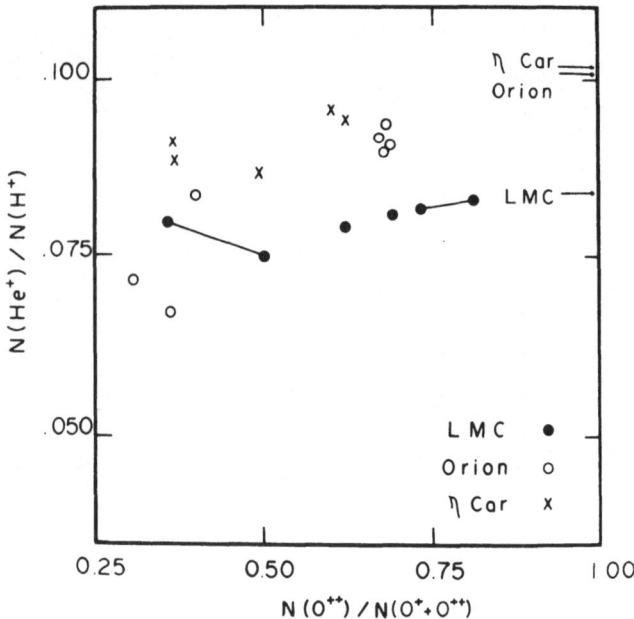

Fig. 4. Comparison of the observed helium ionic abundance
with the observed degree of ionization of oxygen
(Peimbert & Torres-Peimbert, 1974).
The estimate of yo is made from the extrapolated value
at $N(0^{++})/N(0^+ + 0^{++}) = 1$.

To compare then the helium abundance ratio Y to the heavy element abundance Z, it is assumed that oxygen constitutes 45% of Z by mass (Peimbert and Torres-Peimbert, 1974) and thus the oxygen abundance ratio is used as the metallic indicator.

The helium abundance evaluations are made in H II regions presenting very different metallicities, i.e. very different evolutions. These H II regions are either inside our galaxy where the heavy element fraction Z decreases from the center to the outer regions (see Pagel and Edmunds, 1981) or in small irregular and blue compact galaxies presenting large under-abundances in heavy elements (see Lequeux et al., 1979).

The first precise evaluation of primordial helium abundance Y_p extrapolated from several Y over Z observations toward zero Z values is given by Peimbert and Torres-Peimbert (1974). Their deduced result is

$$Y_p = 0.216 \pm 0.02 \quad \text{and} \quad \frac{\Delta Y}{\Delta Z} \sim 3.3.$$

Among other evaluations a reliable one seems to be from Lequeux et al. (1979) who revised previous evaluations and used a larger set of observations toward irregular and blue compact galaxies. Their results shown on Figure 5 yield to

$$Y_p = 0.233 + 0.015 \quad \text{and} \quad \frac{\Delta Y}{\Delta Z} = 1.73 \pm 0.90.$$

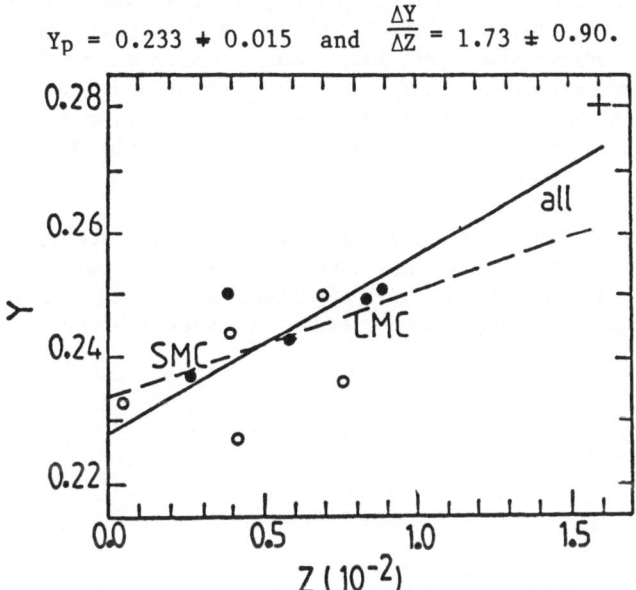

Fig.5. Comparison between observed helium and heavy element abundance by mass. Filled circles are higher quality observations, the cross corresponds to the Orion Nebula. Solid line, the least-squares solution for all objects, dashed line, without Orion. The dashed line is probably more meaningful.

More recent evaluations seem to confirm these results either by
analysing radio and optical recombination lines inside our own
galaxy (Thum, 1981) or by observing the helium gradient decreasing from the center outwards in H II regions inside the M 101
galaxy (Rayo et al., 1982) giving the highest precision result

$$Y_p = 0.216 \pm 0.010.$$

Nevertheless, from other observations the scatter of the Y-Z
diagram is much larger leading to no or more questionable
evaluation of Y_p (French, 1980; Talent, 1980; Kinman and
Davidson, 1981; and Kunth, 1982). In this last case Kunth
(1982) simply deduced an upper limit for Y_p which has to be less
than the average of the observed Y values

$$Y_p < 0.245 \pm 0.003.$$

From these results, all shown on Figure 6, we may say that
the situation of helium observations is rather good. In effect
evaluations made in very different regions and particularly in
different galaxies are quite consistent showing that its
primeval origin is almost certainly settled. Furthermore even
its variation with metallicity seems to be proven, although less
convincingly, giving important results related to the chemical
evolution of galaxies. Particularly, as discussed by Lequeux et
al. (1979), Chiosi and Caimmi (1979), and Chiosi and Matteucci
(1982), it seems that the large $\Delta Y/\Delta Z$ observed ratio could be
explained by assuming heavy mass loss from massive stars. In
effect, this reduces the heavy elements production mainly due to
these massive stars. Nevertheless Maeder (1981) discusses this
assumption.

In conclusion more precise and complete observations are
certainly needed, but because the observational method is almost
unique, one should recall as underlined by Lequeux et al. (1979)
that systematic errors are certainly possible and that they
could affect the Y_p evaluations.

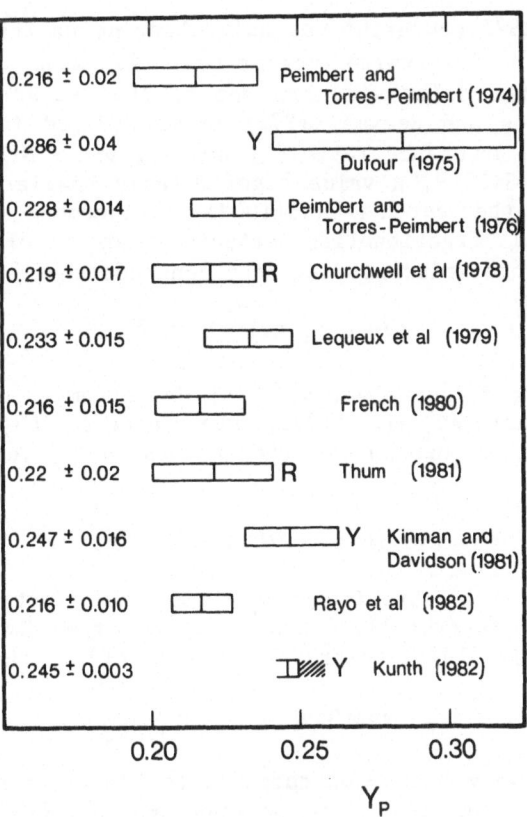

Fig. 6. A summary of evaluations of the primordial
helium abundance Y_p. Measurements noted Y represent
simply helium abundance evaluations without any attempt
to deduce Y_p. We should only have $Y_p < Y$.
Evaluations noted R use radio recombination lines.
From this summary, the reasonable range
for Y_p is $0.20 < Y_p < 0.23$.

IV - OBSERVATION OF INTERSTELLAR DEUTERIUM

Before 1972 deuterium was only observed on the earth either in the ocean or inside meteorites. Assuming that the ^{3}He observed in the solar wind was due to the deuterium burning in the sun, Geiss and Reeves (1972) first deduced indirectly that the deuterium abundance ratio (by number) D/H = N(D)/N(H) should be less than 8.10^{-5}, a value significantly smaller than the one observed on the earth ($\sim 1.5\ 10^{-4}$). Furthermore arguing about some chemical fractionation effects that should take place inside the proto-solar nebula, they conclude that

$$D/H = (2.5 \pm 1)\ 10^{-5}.$$

The first direct observation of deuterium outside the earth came from Beer et al. (1972) who detected CH_3D in Jupiter permitting an evaluation of the D/H ratio by Beer and Taylor (1973) of

$$2.9\ 10^{-5} < D/H < 7.5\ 10^{-5}.$$

But the big surprise came from the first detection of deuterium in interstellar space through the DCN molecule by Jefferts et al. (1973) towards Orion. In effect they found

$$DCN/HCN \sim 3\ 10^{-3},$$

a value extremely high when compared to the observed proto-solar D/H ratio. At the same time Cesarsky et al. (1973) marginally detected the atomic deuterium hyperfine transition at 92 cm toward the galactic center and showed that clearly the D/H value in the interstellar medium should be much smaller

$$3\ 10^{-5} < D/H < 5\ 10^{-4}.$$

This demonstrates that D/H is difficult to deduce from molecular observations since very important chemical fractionation effects are involved. We thus will not discuss here the observations of deuterated molecules in the interstellar space which are certainly less precise than the atomic ones (see a review of these observations in Laurent, 1978). Note here that recent attempts to observe the 92 cm line toward Sgr. A. were made by Sarma and Mohenty (1978) and by Anantharamaiah and Radhakrishnan (1979). They proved with a much better signal to noise ratio than Cesarsky et al. (1973) that this line is undetectable and that D/H toward the galactic center is less than $\sim 6.10^{-5}$, a value still too high to give any important constraints on the galactic evolution models.

Another method for observing directly the atomic species in the far ultraviolet was then proposed by our group which leads to the most direct and precise evaluations. We will present here this observational technique and discuss the results obtained. The idea is to observe directly the far ultraviolet absorption Lyman lines of hydrogen and deuterium with the Copernicus satellite (Rogerson et al., 1973). Due to an isotope shift of ~ 0.3 Å (~ 80 km.s^{-1}) these lines are separated in most interstellar situations since the cloud velocity separations are ~ 10 km.s^{-1}.

Nevertheless, due to the extremely large column density of atomic hydrogen, the deuterium line, lost in the core of the hydrogen line, is almost always unobservable at Lyman α (except in the case of the very nearby cool stars) and often not even observable at Lyman β (see Figure 7, from Vidal-Madjar et al., 1977a).

For these reasons we see that two sets of different obser-vations of the D/H ratio will emerge from these constraints:

- one, less precise, deduced from only Lyman α obser-vations of cool nearby stars;

- the other one from observation of the whole Lyman line series down to Lyman ϵ toward much hotter target stars.

The first precise evaluation of the D/H ratio was given by Rogerson and York (1973) who observed toward β Cen the whole Lyman line series. The result obtained is

$$D/H = 1.25 \, {}^{+\,1.25}_{-\,0.45} \, 10^{-5}$$

demonstrating the precision of the method.$F_S(\lambda)$,

Before giving the results of all the observations made, we should discuss here the basic principles of this approach.

A column density N of atoms along the line of sight will absorb the stellar light $F_S(\lambda)$, giving the final $F(\lambda)$ line profile defined by

$$F(\lambda) = F_S(\lambda) \cdot A(\lambda, N, b, v) \qquad (6)$$

where the function $A(\lambda, N, b, v)$ classically defines the absorption produced by a gas presenting a Maxwellian velocity distribution, i.e.

$$A(\lambda, N, b, v) = \exp \left[- k_0 \, H(a, w) \, N\right]$$

in which H(a,w) is the Harris function describing simulta-
neously the Doppler core of the line and its damping wings.

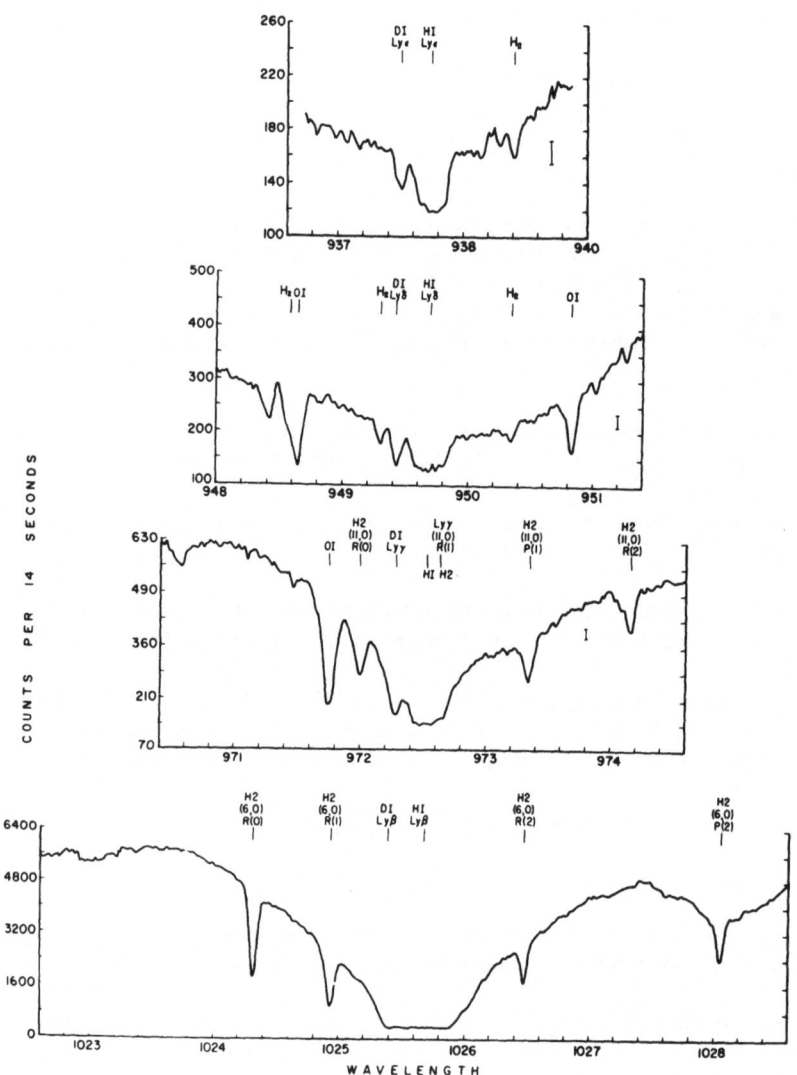

Fig. 7. Observations with Copernicus of four
interstellar Lyman lines toward the star γ Cas.
As labelled, the deuterium lines show up clearly
on the blue wings of the corresponding hydrogen lines.
The stellar Lyman lines are evident as broad absorption
features. Theoretical ± 2 σ error bars appear on the right
for the top three graphs. Detected features are labelled
(from Vidal-Madjar et al., 1977a).

The different parameters are defined as $b = (2\ kT/m)^{\frac{1}{2}}$, the broadening parameters, $a = 1/t.\Delta v_D$ where t is the lifetime of the transition and $\Delta vD = v_0.b/c$, $w = \Delta v/\Delta v_D$ represents the frequency separation from line center,

$$k_0 = \frac{\pi e^2}{mc}\ f\ \frac{1}{\sqrt{\pi}}\ \frac{1}{\Delta v_D}$$

is the absorption coefficient per atom at the center of the line and f is the oscillator strength of the transition.

From relation (6) it is possible to calculate the equivalent width of an absorption line

$$W_\lambda = \int\ [1 - A(\lambda,N,b,v)]\ d\lambda \qquad (7)$$

leading to the well known curve of growth relating W_λ/λ to the product $N\ f\ \lambda$.

Although abundance evaluations are relatively simple through curve of growth analysis, the precision of determinations corresponding to equivalent width falling on the flat part of the curve of growth can only be rather poor. Furthermore several additional uncertainties can arise from either the unknown stellar profile $F_S(\lambda)$ or by the fact that different absorbing clouds are often present on the line of sight at different radial velocities (as demonstrated by Gomez-Gonzalez and Lequeux, 1975).

For these reasons it was necessary to develop line profile analysis to use the information contained not only in the equivalent width of the line but also in its shape as explained by Vidal-Madjar et al. (1977a).

If in the case of the observations toward nearby cool stars the knowledge of the stellar profile $F_S(\lambda)$ presents a serious difficulty, in the case of remote stars this problem could be reduced by selecting hot stars presenting the highest possible rotational velocity v sini. Thus although unknown, $F_S(\lambda)$ has to be smooth over a large spectral range, not affecting then very much the deuterium abundance evaluation.

The results obtained are presented on Figure 8 for the observation toward nearby cool stars and on Figure 9 toward more distant O and early B stars.

As it is quite obvious we are not at all in a situation similar to the one related to the helium observations. In both cases, the scatter of the observed values is quite large and

seems to reach a factor of 10. Although it is already
surprising to see such variations within ~ 1000 pc from the sun
(Figure 9), this looks unbelievable within only 30 pc from the
sun (Figure 8).

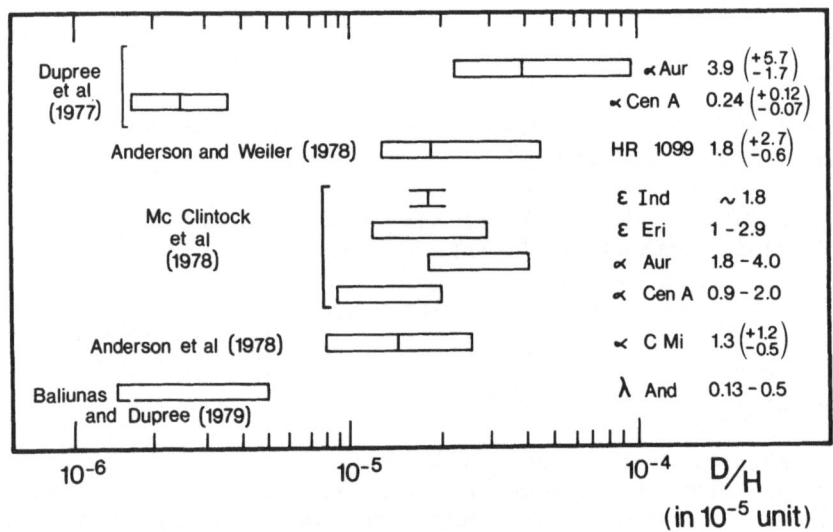

Fig. 8. D/H evaluations made through the study
of only the Lyman α line profile observed
toward nearby (d < 30 pc) cool stars.

Either the cosmological significance of deuterium is wrong
and this element is formed in yet unknown sites, or mechanisms
exist able to segregate efficiently deuterium in the interstellar space.

Search for segregating mechanisms was first attempted by
Vidal-Madjar et al. (1977b, 1978) who argued that a selective
radiation pressure could act on deuterium atoms and not on
hydrogen ones (the hydrogen atoms being self shielded from the
Lyman lines flux coming from the stars) producing some spatial
separation of the two species. Because they underlined that the
solar system was in a very peculiar position, i.e. embedded at
the edge of a small interstellar cloud (a situation widely
confirmed now through e.g. hydrogen column densities evaluated
toward white dwarfs), this mechanism could produce the observed
D/H variations.

Bruston et al. (1981) reanalysed in more detail this
mechanism showing that it should produce deuterium abundance
changes between two regions presenting a temperature gradient
rather than a density gradient as initially suggested by Vidal-
Madjar et al. (1977b, 1978).

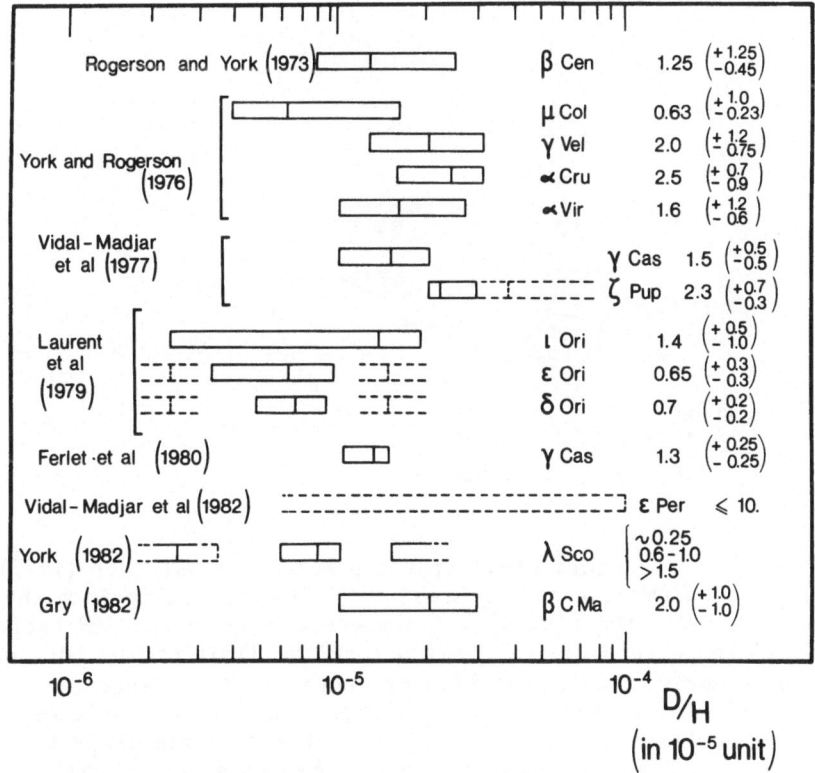

Fig. 9. D/H deduced from observations of several
Lyman lines toward early type stars (d ≤ 1000 pc).

In particular they show that without any ionization effects
one should have

$$\left(\frac{D}{H}\right)^b = \left(\frac{D}{H}\right)^a \left(\frac{T^b}{T^a}\right)^{\frac{1}{2}} \qquad (8)$$

where subscript a refers to the region at temperature T^a and b
to the one at temperature T^b. If ionization effects are added
they show that

$$\left(\frac{D}{H}\right)^b = \left(\frac{D}{H}\right)^a \left[\left(\frac{T^b}{T^a}\right)^{\frac{1}{2}} + x^b - x^a \left(\frac{T^b}{T^a}\right)^{\frac{1}{2}} \right] \qquad (9)$$

where x^a and x^b are the hydrogen or deuterium ionization rates
in regions a and b. From these relations and using a model of
interstellar cloud taken from McKee and Ostriker (1977) they
evaluate (see Figure 10) that depleted regions by approximately
a factor 2 and enriched regions by a factor 10 should exist in
some interstellar clouds.

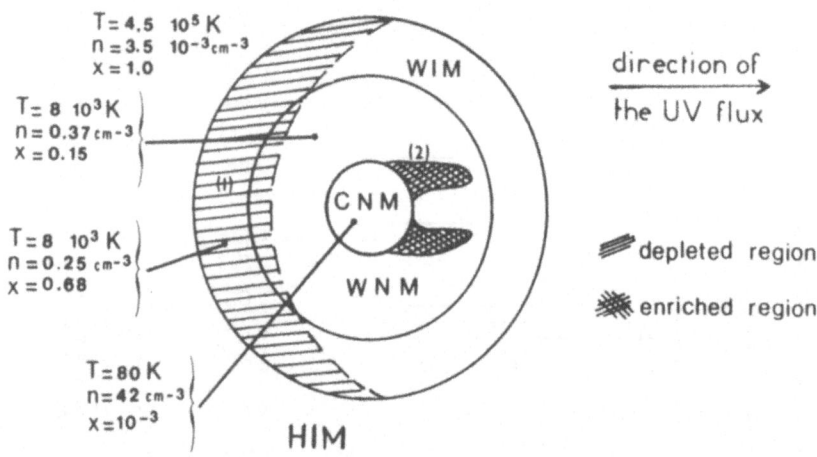

Fig. 10. A standard cloud from the McKee and Ostriker (1977)
model. Assuming that the stellar UV flux is coming from the
left (arrow), two regions are represented where the D/H ratio
should be perturbed: (i) region 1 at the limit of the hot and
warm medium where depletion of deuterium is expected, and
(ii) region 2, behind the cold core where strong enrichment
in deuterium is predicted. Notice the probable shape of
region 2 due to UV shielding by the dense core itself,
stopping on its axis the radiation pressure mechanism
(from Bruston et al., 1981).

This type of mechanism could explain the spread of observed
D/H values in both cases, and because the depleted deuterium
region has to be more extended than the enriched one, Bruston
et al. (1981) conclude that the unperturbed D/H value of the
interstellar medium should be more probably a high value, i.e.
of the order of $2.25 \ 10^{-5}$.

Other segregation effects have been also considered.
Bruston et al. (1981) discuss the possibility of apparent atomic
deuterium enrichment (versus atomic hydrogen) at the edge of
interstellar clouds containing H_2 and HD molecules (due to the
self-shielding of H_2 against photo-dissociation. This mechanism
seems possible (see e.g. Watson, 1974) but not directly
applicable to the observations presented in Figures 8 and 9
which correspond to lines of sight with too small amounts of H_2
and DH molecules (see Spitzer and Morton, 1976).

Another interesting study made by York and Jura (1982)
shows a possible weak trend between deuterium and zinc

abundances. Surprisingly a correlation seems to exist which may indicate that the variations, if real, are not due to galactic evolution effects (they should then be anti-correlated as the observations of deuterated molecules toward the galactic center seem to indicate). This means that either deuterium is not cosmological (a remote possibility) or that they could both suffer a common depletion mechanism. This result is nevertheless still marginally convincing and needs further studies to be confirmed.

The matter seemed to be settled until the more recent evaluation made by Vidal-Madjar et al. (1982a) toward ε Per.

The deuterium Lyman γ and δ lines toward ε Per present a very surprising peculiarity: the weakest line is stronger than the stronger one ! How can this be ? Using several arguments and particularly because high velocity H I gas is seen on the blue wings of the Lyman lines (probably due to H I material in the ε Per stellar wind) they conclude that the deuterium lines are probably blended with an H I component moving away from the star at approximately 80 km.s^{-1}. Because the observations of the two deuterium lines are not simultaneous, it is thus acceptable to say that this component changed in column density by more than a factor 3 between the two observations (separated by roughly 10 hours) explaining the difference observed.

From a suggestion made recently by M. Jura, it was possible to check directly this idea by separating the observational data into several sets in order to see directly if during the observations of a given deuterium line, its shape was changing. This was done by dividing the data into two sets in the case of each line (see Figures 11 and 12).

The result obtained seems to indicate that not only time variations are indeed present over the deuterium lines but also in the different high velocity H I gas lines. This is logical since, according to the Vidal-Madjar et al. (1982a) analysis, all these lines are probably due to H I material in the stellar wind. This last result also confirms the recent study of Gry et al. (1982) which revealed the presence of several weak lines probably due to high velocity H I gas in the stellar winds of early-type stars.

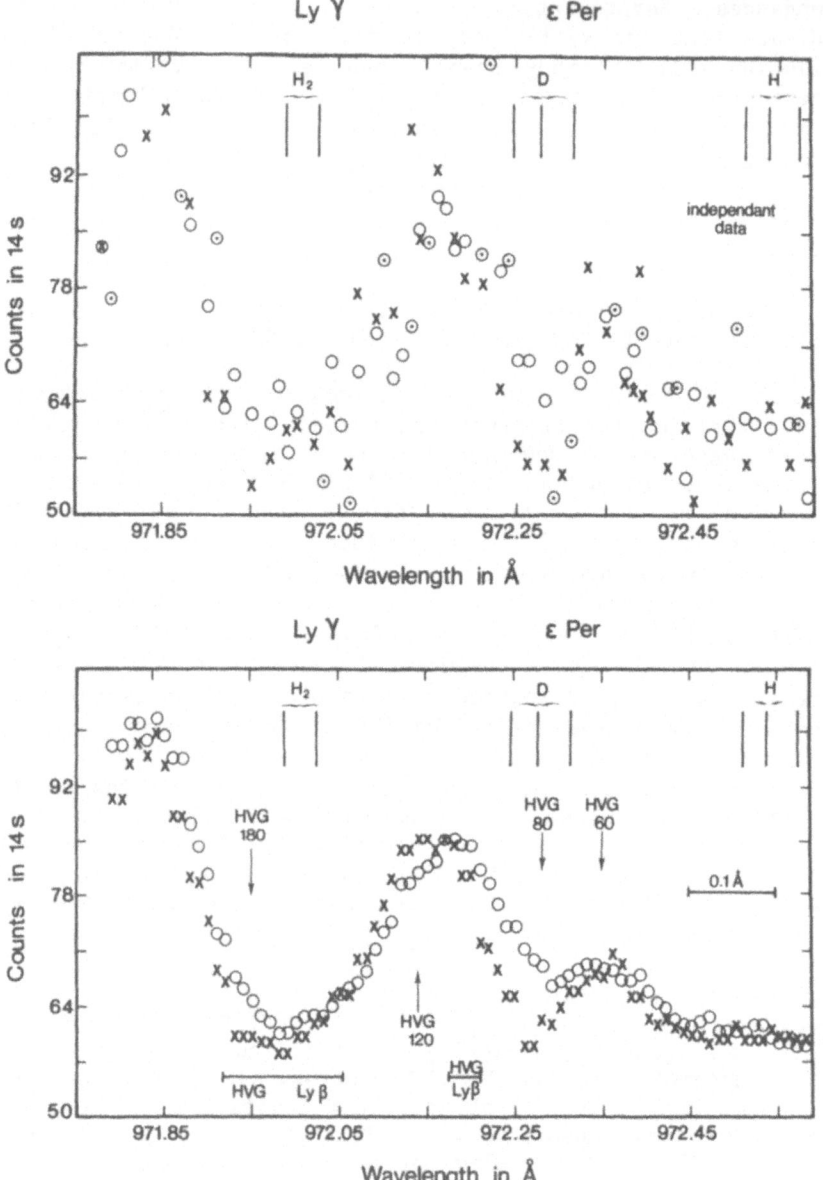

Fig. 11. Observations are separated into two sets:
crosses (X) for the first one, circles (o) for the second one.
The upper part shows independent data points averaged
over 0.01 Å intervals (points with tiny dots away
from the others have a very low weight < 0.2).
The lower part shows the same measurements averaged
over a sliding interval of 0.1 Å.   Arrows indicate high
velocity gas (HVG) observed in the two Lyman lines.
Horizontal bars indicate HVG detected near Lyman β.

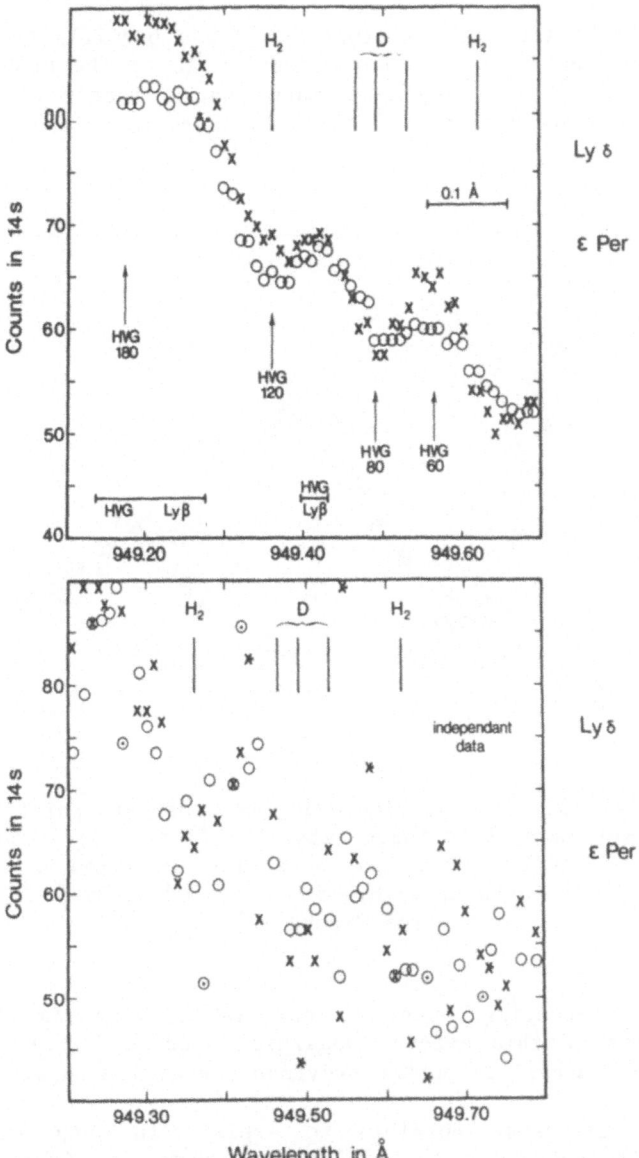

Fig. 12. Same as Figure 11 but for Lyman δ.

To check further this hypothesis, we divided the Lyman γ data into three sets in order to observe further fluctuations. The result is shown on Figure 13, revealing that the time fluctuation is more complicated than a simple line weakening over the deuterium feature. In effect it seems that, while the line is weakening on one side, a build up appears on the other side, like if the hydrogen atoms were slowly drifting in velocity.

By simply fitting the profiles it was possible to evaluate that the needed hydrogen column density is of the order of few times 10^{14} cm^{-2} and changed by more than a factor of 6 within one hour (a whole data set is taken in approximately 3 hours).

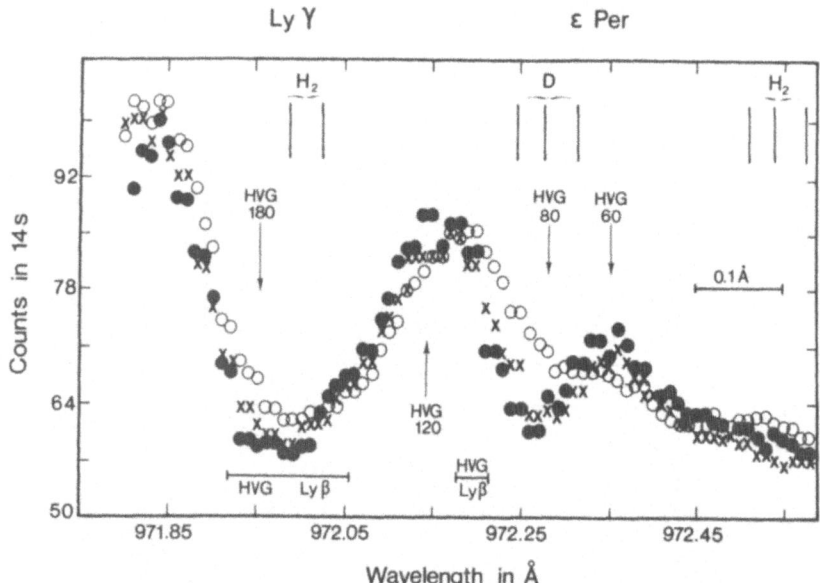

Fig. 13. Same as Figure 11, but the data points are separated into three sets: (X) first, (o) second, and (o) third. Note that some time variation is now also evident at HVG-60, detected at Ly δ (see Figure 12).

The interesting question now is to try to draw the consequences of this effect (possible blend of D lines by high velocity H I material) on the previous D/H evaluations.

First it seems possible to explain in such a way the scatter observed among the evaluations shown on Figure 9. The effect being an apparent increase in deuterium abundance we may conclude that the real D/H value in the interstellar medium is more probably the lowest observed one, i.e.

$$D/H \sim 2.5 \ 10^{-6}.$$

But because the mechanism proposed by Bruston et al. (1981) produces opposite consequences, and because we do not know if this blending of the line often occurs, we are in fact unable to conclude until we understand in a better way the physics of this blending mechanism.

This was discussed by Vidal-Madjar et al. (1982b) who
propose that the blending could be due to a line locking
mechanism (see e.g. Scargle, 1973) which stops the acceleration
of the stellar wind at given velocities, the H I atoms being
simply then tracers of the wind condensations.   Also, as
suggested by Lamers (1982), these features could be due to inhomogeneities at or near the stellar surface, like spots or
prominences.   In any one of these cases, it is clear that the
blending effect should be a function of the star's evolution and
particularly a function of the class of luminosity of the star.
This is why, as shown on Figure 14, Vidal-Madjar et al. (1982b)
looked for a correlation between D/H and the luminosity class of
the target star.

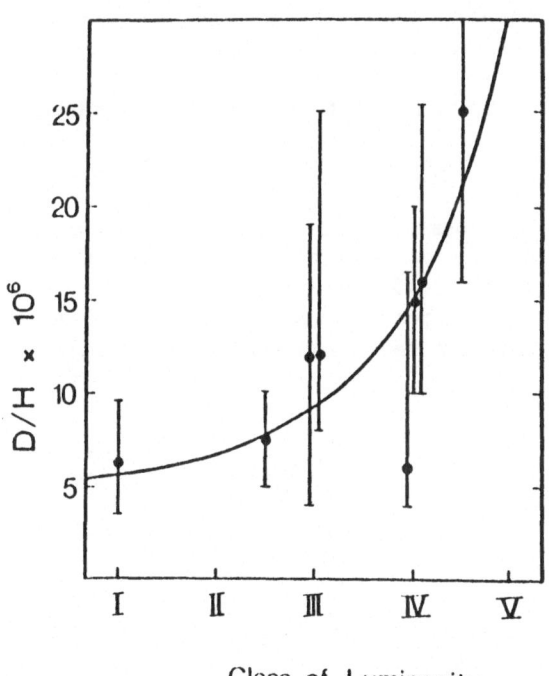

Class of Luminosity

Fig. 14. Correlation of the D/H evaluations
with the stellar class of luminosity: class V are main
sequence stars while class I are more evolved supergiants
(from Vidal-Madjar et al., 1982b).

A trend is obviously observed, showing that a correlation
seems to exist between the D/H evaluations and the stellar class
of luminosity.   This suggests finally that the blending
mechanism seems to be the dominant one (the mechanism proposed
by Bruston et al. (1981) obviously cannot explain such a

trend). Note that the two low values observed by Laurent et al. (1979) represent a strong constraint since they correspond to an average over three components present on these lines of sight. This gives more weight to these D/H evaluations when compared to the others. If confirmed this indicates that the D/H abundance ratio in the interstellar medium near the sun is rather

$$D/H \sim 5.10^{-6} \ (\pm \ 3.10^{-6}).$$

It is not possible for the moment to observe less evolved regions as in the case of the helium study. Thus the evaluation of primordial deuterium is not straightforward. The trick used here is to compare this value to the one observed inside the solar system in the giant planet (see e.g. Kunde et al., 1982; Encrenaz and Combes, 1982), i.e. to the D/H value that was present in the galaxy $4.5.10^9$ years ago.

Because observations in the solar system range around

$$D/H \sim 3.10^{-5} \ (\pm \ 1.10^{-5})$$

it seems that the cosmological origin and subsequent destruction of deuterium is more or less confirmed. Nevertheless much work is needed and particularly if these last values are proved to be correct, then galactic evolution models should be revised since they predict a smaller deuterium abundance variation.

V - CONCLUSION

The set of helium and deuterium observations seem to confirm the cosmological origin of both elements. Nevertheless the scatter of observational points is still quite large and may indicate either underestimated observational difficulties or segregation mechanism still not fully understood.

It seems possible to evaluate the primordial values of both species: for helium in a straightforward manner by observing regions presenting very different evolutions, and for deuterium either through models of galactic evolution or through the comparison between interstellar and planetary evaluations.

We may conclude from Figure 6 that

$$0.20 < Y_p < 0.23$$

while for deuterium, following Pagel (1982), we may say that the galactic evolution should decrease D/H by a factor of 2, or that the deuterium abundance by mass X_D compared to the primordial one X_{DP} should be

$$0.37 < \frac{X_D}{X_{DP}} < 0.73$$

showing that we should have if D/H $\sim 5.10^{-6}$

$$1.1 \ 10^{-5} < X_{DP} < 2.1 \ 10^{-5}.$$

If on the contrary we ignore the galactic evolution models and say that in $4.5 \ 10^9$ years the D/H ratio decreased by a factor 6 we may roughly extrapolate to the galactic birth, i.e. to the primordial value, by assuming a change of a factor 12. This gives

$$X_{DP'} \sim 8.5 \ 10^{-5}.$$

Recently Spite and Spite (1982) evaluated the primordial ^7Li abundance to be Li/H = 1.10^{-10} or X_{Li} = $5.96 \ 10^{-10}$ and placed new constraints on the standard Big Bang model. Following their approach (see Figure 15) we can see that if we use the deuterium X_{DP} evaluation, the standard Big Bang seems to be in difficulty. If $X_{DP'}$ is used, the conflict is less severe with the Y_p estimation but still it is difficult to fit all estimations together.

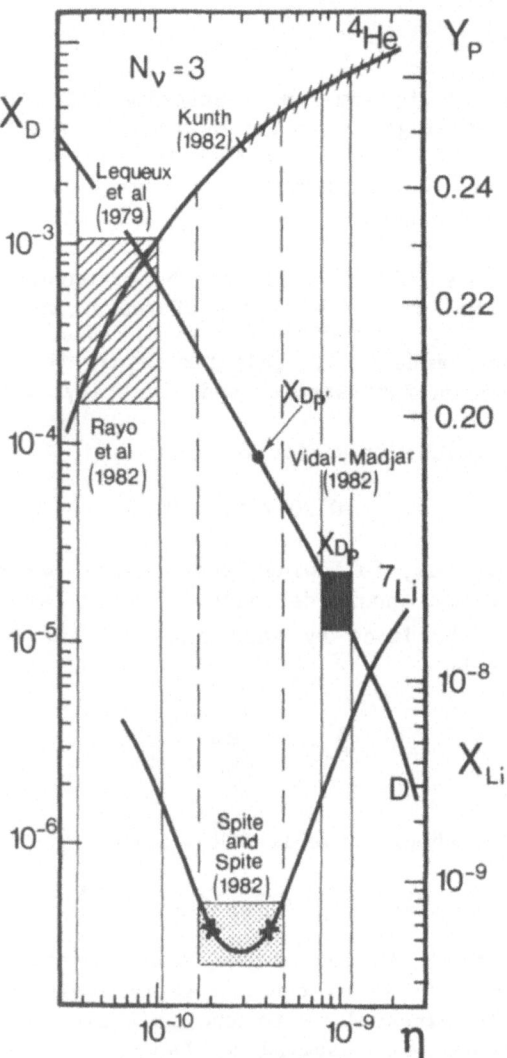

Fig. 15. Theoretical calculations from Schramm (1982)
of Y_p, X_D and X_{Li} versus the baryons to photons ratio.
The boxes indicate the observational evaluations
along with the "compatibility zone.
Helium and deuterium estimations seem to be incompatible
in the frame of the standard Bib Bang model.

Before trying to reject the standard Big Bang model, it is important to recall that still many errors are inherent to these estimations. More observations are thus eagerly needed.

What type of observations ?

For helium, other observational techniques should be added to the certainly needed continuation of high quality observations of emission nebulae. A possibility is to observe He I and He II in the nearby interstellar medium, in absorption in the far UV, in the direction of a hot white dwarfs. This observation will give an absolute point on the Y-Z diagram independent of possible systematic errors inherent to the other observational technique. This could well shift the whole set of points and induce a different primordial Y_p. Another possibility is the direct observation of helium entering the solar system. Although difficult this observation represents also an independent approach (see e.g. Meier, 1981).

For deuterium, the extrapolation toward the primordial values is very hazardous and observation of sites presenting different evolutions is absolutely necessary. Two possibilities exist: i) to observe directly in the far UV the interstellar medium much further away in our own galaxy or even in the less evolved galaxies that are the Magellanic Clouds in order to start to build, like in the helium case, a X_D-Z diagram, or (as suggested by Adams, 1976, and Bahcall, 1979) ii) to detect in front of quasars the deuterium line corres-ponding to one of the primordial H I clouds recently discovered by Sargent et al. (1981). A complete parametric study of this observational possibility was developed by Laurent and Vidal-Madjar (1981).

Although the cosmological implication of these observations is very important, one should not forget the important consequences that they will probably have on our understanding of galactic evolution. The past has proven that we can expect a tremendous number of surprises in the future.

Note added

A well known theoretical prediction (Bodenheimer, 1966) indicates that for stars having masses larger than 3 M_o, no convective mixing will bring the upper layers of the star towards its interior. This proves that deuterium should survive at least some time at the surface of the large mass stars. If true, this should yield to a new observational technique to evaluate the D/H ratio. But due to mass loss and to the star evolution, after some time this deuterium should burn out, a fact demonstrated by the Peimbert et al. (1981) observation of

Canopus (an evolved large mass star: F0 Ib type). In effect they found that in the star atmosphere the D/H ratio is less than $9.0 \ 10^{-6}$. Recently Ferlet et al. (1982) evaluated that in Canopus the D/H ratio was even less than $5.5 \ 10^{-7}$ proving clearly that deuterium was burned out in the star atmosphere.

These evaluations of the D/H ratio in stellar atmospheres through the study of the Balmer α line profiles may prove to be in the future interesting and could represent a new observational technique to be added to the one listed herein.

ACKNOWLEDGEMENTS

Extremely interesting last minute corrections are certainly due to H.J.G.L.M. Lamers, J. Audouze and S. Vauclair that I am pleased to thank here, but also to the whole spirit given at the Cargese Summer School by M.F. Hanseler, M. Levy and J.L. Basdevant, the organisers, the lecturers and the participants that I want to warmly acknowledge here.

It is a pleasure to thank M. Jura who made a very fruitful suggestion concerning the study of time variations in deuterium lines. I am also indebted to R. Ferlet, C. Gry and D.G. York who communicated to me unpublished material. It is a pleasure to thank them here. I am also very grateful to J. Lequeux and C. Laurent who made many constructive comments on this manuscript.

The bibliographic preparation of this work was made very easy owing to F. Marchand who merits many thanks along with N. Giraud who edited this document.

REFERENCES

Adams, T.F.,   1976,   A. & A., 50, 461.
Alpher, R.A., Bethe, H.A., and Gamov, G, 1948, Phys. Rev., 73, 803.
Anantharamaiah, K.R. and Radhakrishnan, V, 1979, A.& A. Letters, 79, L9.
Anderson, R.C., Henry, R.C., Moos, H.W., and Linsky, J.L., 1978, Ap. J., 226, 883.
Anderson, R.C. and Weiler, E.J., 1978, Ap. J., 224, 143.
Anglin, J.D., 1975, Ap. J., 198, 733.
Anglin, J.D., Dietrich, W.F. and Simpson, J.A., 1973, Ap. J. Letters, 186, L41.
Audouze, J. and Tinsley, B.M., 1974, Ap. J., 192, 487.
Audouze, J. and Tinsley, B.M., 1976, Ann. Rev. Astron. Ap., 14, 43.
Audouze, J. and Vauclair, S., 1980, An introduction to nuclear astrophysics, D. Reidel Publishing Company, Dordrecht.
Austin, S.M., 1980, Prog. in Port. and Nucl. Phys.
Bahcall, J.N., 1979, IAU Colloquium No. 54, NASA-CP-2111, 215.
Baliunas, S.L. and Dupree, A.K., 1979, Ap. J., 227, 870.
Beaudet, A. and Yakil, A., 1977, Ap. J., 218, 253.
Beckers, J.M., 1975, Ap. J. Letters, 195, L43.
Beer, R., Farmer, C.B., Norton, R.H., Martonchik, J.V. and Barnes, T.G., 1972, Science, 175, 1360.
Beer, R. and Taylor, F.W., 1973, Ap. J., 179, 309.
Bodenheimer, P., 1966, Ap. J., 144, 103.
Brocklehurst, M., 1971, M.N.R.A.S., 153, 471.
Brocklehurst, M., 1972, M.N.R.A.S., 157, 211.
Bruston, P., Audouze, J., Vidal-Madjar, A. and Laurent C., 1981, Ap. J., 243, 161.
Cesarsky, D.A., Moffet, A.T. and Pasachoff, J.M., 1973, Ap. J. Letters, 180, L1.
Chiosi, C. and Caimmi, R., 1979, A. & A., 80, 234.
Chiosi, C. and Matteucci, F.M., 1982, A. & A., 105, 140.
Chiosi, C., Nasi, E., and Sreenivasan, S.R., 1978, A. & A., 63, 103.
Chupp, E.L., Forrest, D.J., Higbie, P.R., Suri, A.N., Tsai, C. and Dunphy, P.P., 1973, Nature, 241, 333.
Churchwell, E., Smith, L.F., Mathis, J., Mezger, P.G. and Hutchmeier, W., 1978, A. & A., 70, 719.
Coleman, G.D. and Worden, S.P., 1976, Ap. J., 205, 475.
Colgate, S.A., 1973, Ap. J. Letters, 181, L53.
David, Y. and Reeves, H., 1980, Physical Cosmology, Les Houches, Session XXXII, Balian, R., Audouze, J. and Schramm, D.N., editors, North Holland Publishing Company, 443.
Dufour, R.J., 1975, Ap. J., 195, 315.
Dupree, A.K., Baliunas, S.L. and Shipman, H.L., 1977, Ap. J., 218, 361.

Encrenaz, T. and Combes, M., 1982, preprint.
Epstein, R.I., 1977, Ap. J., 212, 595.
Epstein, R.I., Lattimere, J.M. and Schramm, D.N., 1976,
    Nature, 263, 198.
Ferlet, R., Dennefeld, M., Laurent, C. and Vidal-Madjar, A.,
    1982, A. & A., submitted.
Ferlet, R., Vidal-Madjar, A., Laurent, C. and York, D.G.,
    1980, Ap. J., 242, 576.
French, H.B., 1980, Ap. J., 240, 41.
Geiss, J. and Reeves, H., 1972, A. & A., 18, 126.
Gomez-Gonzalez, J. and Lequeux, J., 1975, A. & A., 38, 29.
Gry, C., 1982, private communication.
Gry, C., Vidal-Madjar, A., Bruston, P., Laurent, C. and
    York, D.G., 1982, Ap. J., submitted.
Hayashi, C., 1950, Progr. Theor. Phys., 5, 224.
Hoyle, F. and Fowler, W.A., 1973, Nature, 241, 384.
Jefferts, K.B., Penzias, A.A. and Wilson, R.W., 1973,
    Ap. J. Letters, 179, L57.
Kinman, T.D. and Davidson, K., 1981, Ap. J., 243, 127.
Kunde, V., Hanel, R., Maguire, W., Gautier, D., Baluteau, J.P.,
    Marten, A., Chedin, A., Husson, N. and Scott, N., 1982,
    Ap. J., in press.
Kunth, D., 1982, Ph. D. Thesis, Paris University.
Lamers, H.J.G.L.M., 1982, private communication.
Lattimer, J. and Schramm, D.N., 1974, Ap. J. Letters,
    192, L45.
Laurent, C., 1978, Ph. D Thesis, University of Paris VII.
Laurent, C., Vidal-Madjar, A. and York, D.G., 1979, Ap. J.,
    229, 923.
Laurent, C. and Vidal-Madjar, A., 1981, Haute Résolution
    Spectrale en Astrophysique, Deuxième Colloque National du
    Conseil Français du Télescope Spatial, p. 145, Orsay.
Lequeux, J., Peimbert, M., Rayo, J.F., Serrano, A. and Torres-
    Peimbert, S., 1979, A. & A., 80, 155.
Maeder, A., 1981, A. & A., 101, 385.
McClintock, W., Henry, R.C., Linsky, J.L. and Moos, H.W.,
    1978, Ap. J., 225, 465.
McKee, C.F., and Ostriker, J.P., 1977, Ap. J., 218, 148.
Meier, R.R., 1981, A Parametric Study of Interstellar Helium
    Atoms Incident upon the Earth, NRL Memorandum report 4423.
Meneguzzi, M., Audouze, J. and Reeves, H., 1971, A. & A.,
    15, 337.
Osterbrock, D.E., 1974, Astrophysics of Gaseous Nebulae,
    W.H. Freeman and Company.
Pagel, B.E.J., 1982, Abundance of elements of cosmological
    interest. Presented at the Royal Society.
Pagel, B.E.J. and Edmunds, M.G., 1981, Ann. Rev. Astron.
    Astrophys., 19, 77.
Peebles, P.J.E., 1971, Physical Cosmology, Princeton University
    Press.

Peimbert, M., Rodriguez, L.F., and Torres-Peimbert, S., 1974,
     Rev. Mexicana Astron. Astrof., 1, 129.
Peimbert, M. and Torres-Peimbert, S., 1974, Ap. J., 193, 327.
Peimbert, M. and Torres-Peimbert, S., 1976, Ap. J., 203, 581.
Peimbert, M. and Torres-Peimbert, S., 1977, M.N.R.A.S.,
     179, 217.
Peimbert, M., Wallerstein, G. and Pilackowski, C.A., 1981,
     A. & A., 104, 72.
Rayo, J., Peimbert, M. and Torres-Peimbert, S., 1982, Ap. J..
Reeves, H., 1974, Ann. Rev. Astron. Astrophys., 12, 437.
Reeves, H., Audouze, J., Fowler, W.A. and Schramm, D.N., 1973,
     Ap. J., 179, 909.
Reeves, H. and Meyer, J.P., 1978, Ap. J., 226, 613.
Rogerson, J.B., Spitzer, L., Drake, J.F., Dressler, K.,
     Jenkins, E.B., Morton, D.C. and York, D.G., 1973, Ap. J.
     Letters, 181, L97.
Rogerson, J.B. and York, D.G., 1973, Ap. J. Letters, 186, L95.
Sargent, W.L.M., Young, P.J., Boksenberg, A. and Tytler, D.,
     1980, Ap. J. Suppl., 42, 41.
Sarma, N.V.G. and Mohanty, D.K., 1978, M.N.R.A.S., 184, 181.
Scargle, J.D., 1973, Ap. J., 179, 705.
Schramm, D.N., 1982, Phil. Trans. R. Soc. A., in press.
Schramm, D.N. and Stiegman, G., 1979, Phys. Lett., 87B, 141.
Schramm, D.N. and Wagoner, R.V., 1977, Ann. Rev. Nucl. Sci.,
     27, 37.
Sciama, D.W., 1971, Modern Cosmology, Cambridge University
     Press.
Searle, L. and Sargent, W.L.W., 1972, Ap. J., 173, 25.
Seaton, M.J., 1960, Rep. on Prog. in Phys., 23, 313.
Spite, F. and Spite, M., 1982, preprint.
Spitzer, L. Jr. and Morton, W.A., 1976, Ap. J., 204, 731.
Stiegman, G., 1979, Ann. Rev. Nucl. Part. Sci., 29, 313.
Stiegman, G., Schramm, D.N. and Gunn, J.E., 1977, Phys. Lett.,
     66B, 202.
Talent, D.L., 1980, Ph. D Thesis, Nice University.
Tayler, R.J., 1980, Rep. Prog. Physics, 43, 253.
Thum, C., 1981, Vistas in Astr., 24, 355.
Tinsley, B.M., 1980, Fund. Cosmic. Phys., 5, 287.
Truran, J.W. and Cameron, A.G.W., 1971, Ap. Space Sci.,
     14, 179.
Vidal-Madjar, A., Audouze, J., Bruston, P. and Laurent, C.,
     1977b, La Recherche, 8, 617.
Vidal-Madjar, A., Ferlet, R., Laurent, C. and York, D.G.,
     1982a, Ap. J., 260, 128.
Vidal-Madjar, A., Laurent, C., Bonnet, R.M. and York, D.G.,
     1977a, Ap. J., 211, 91.
Vidal-Madjar, A., Laurent, C., Bruston, P. and Audouze, J.,
     1978, Ap. J., 223, 589.
Vidal-Madjar, A., Laurent, C., Gry, C., Bruston, P.,
     Ferlet, R., York, D.G., 1982b, A. & A., submitted.

Vigroux, L., 1979, Ph. D. Thesis, University of Paris South.
Vigroux, L., Audouze, J. and Lequeux J., 1976, A. & A., 52, 1.
Wagoner, R.V., 1969, Ap. J. Suppl., 18, 147.
Wagoner, R.V., 1973, Ap. J., 179, 343.
Watson, W.D., 1974, Les Houches, Session XXVI, North Holland
     Publishing Company.
Whitford, A.E., 1958, Astron. J., 63, 201.
Yang, J., Schramm, D.N., Stiegman, G. and Rood, R.T., 1979,
     Ap. J., 227, 697.
York, D.G., 1982, Ap. J., in press.
York, D.G. and Jura, M., 1982, Ap. J., 254, 88.
York, D.G. and Rogerson, J.B., 1976, Ap. J., 203, 378.

NUCLEOSYNTHESIS AND CHEMICAL EVOLUTION OF GALAXIES

Jean Audouze

Institut d'Astrophysique, 98bis Bld Arago, F-75014
Paris

Abstract : This chapter deals with a brief presentation of the
nucleosynthesis processes which are responsible for the formation
of the observed chemical elements : the primordial nucleosynthesis occurring just after the Big Bang, responsible for the formation of D, ^{3}He, ^{4}He and at least in part ^{7}Li ; the bombardment
of the interstellar medium by galactic cosmic rays responsible
for the formation of the bulk of the Li, Be, B elements ; and the
stellar nucleosynthesis responsible for the formation of the heavier nuclear species.
      The abundances of these elements evolve with time and can be
different from an astrophysical site to another. The principles
which govern the chemical evolution of galaxies are stated herein, and the recent works concerning this problem are mentioned.
The stochastic metal enrichment in galaxies, the evolution of
the ^{7}Li abundance with time, the effect of the infall of external
gas on the evolution of the galactic disk, the use of planetary
nebulae to try to deduce the rate of star formation for low mass
stars etc... are reviewed. Finally the future of the studies concerning the nucleosynthesis and the chemical evolution galaxies
is discussed.

1 - Introduction.

The purpose of this presentation is to bring the reader
up to date on some of the most recent developments concerning
the nucleosynthetic processes (which transform the chemical
composition of the observed matter in the Universe) such as the
models which attempt to describe the evolution with time, of this
composition observed in different astrophysical sites.

The abundance N_i of a given chemical element i is function

95

J. Audouze et al. (eds.), Diffuse Matter in Galaxies, 95–140.
Copyright © 1983 by D. Reidel Publishing Company.

of (i), the location r, (ii), the age t of the astrophysical object or the medium in which the abundance is observed or determined, and (iii), the abundances of other chemical elements j ;

$$N_i = N_i \ (r,t,N_i) \hspace{6cm} (1)$$

Our task is therefore to provide some information on such functions as (1), namely, how the observed abundances depend on the above mentioned variables. Although the final solution to such a formidable problem is not in view, some progress has been made toward the determination of solutions for significant subsets of the Mendeleev atomic tables.

The organization of this presentation is as follows: Section 2 is devoted to a brief updated survey of the nucleosynthetic processes responsible for the formation of the observed chemical species. Section 3 provides the basic principles governing the chemical evolution of galaxies while Section 4 reviews some recent developments concerning different problems relevant to this type of evolution. Among them are a preliminary study of stochastic enrichment of galaxies, the evolution of ^{7}Li during the galactic history, the effect of infall of external gas on the evolution of the galactic disk, the use of planetary nebulae to attempt to determine the rate of star formation in the low mass (M $\leqslant$ 3-5M$_\odot$) range etc... Finally Section 5 contains some conclusions concerning the present status of these problems.

2 - A quick survey of the nucleosynthetic processes.

The nucleosynthetic processes responsible for the formation of the observed chemical species occur either during the primordial phases of the Universe, or on the galactic scale, i.e. in the interstellar medium itself or in stars. The primordial nucleosynthesis is responsible for the formation of the light elements (D, ^{3}He, ^{4}He) ; the bombardment of the interstellar medium by the galactic cosmic rays is responsible for the formation of ^{6}Li, ^{9}Be, ^{10}B and ^{11}B (the case of ^{7}Li is explained later) while the heavier elements are produced by the stars either during the "normal" course of their evolution or at the end of their lives when stars explode as supernovae.

Due to the restricted format of this presentation there is no room to present a full account of all the aspects of these three nucleosynthetic processes. Interested readers are referred to [1, 2, 3 and 4] for further developments.

2 - 1 The primordial nucleosynthesis.

Many recent articles (see eg. [5] and [6]) review the basic aspects of the primordial nucleosynthesis. I would like to summarize here the currently accepted views regarding this problem, such as the impact of the recent Li abundance determination performed by Spite and Spite [7] concerning some old halo stars.

2 - 1. 1 <u>The relevant D, ^{3}He, ^{4}He and ^{7}Li abundances.</u>

The chapter written by A. Vidal-Madjar (this book) describes the observational material concerning D and ^{4}He. I will adopt here X(D) = 5 10^{-5} (with a large error bar due to the dispersion of the interstellar and the Solar System abundance determinations). For ^{4}He Kunth [8] sets an upper limit Y_p < 0.25 and I would adopt here Y_p = 0.23 ± 0.01.

Concerning ^{3}He an upper limit of ^{3}He/H < 5 10^{-5} has been set up by [9]. A figure of X(^{3}He) = 3 10^{-5} will be used here.

In the case of ^{7}Li, Spite and Spite [7] have observed very recently with the CFH 3.60 m telescope about a dozen faint halo stars with effective temperatures T_{eff} ranging from 5100 to 6300 K. For these stars with $T_{eff} \gtrsim$ 5500 K they found the same Li abundance such that Li/H $\approx 10^{-10}$. It is interesting to note that in the same effective temperature range the Li/H abundance decreases with T_{eff} in the disk stars from about 10^{-9} for the hottest stars to less than 10^{-10} for the coldest ones. From such observations, Spite and Spite [7] derive two very interesting consequences : (i) the primordial ^{7}Li abundance is such that ^{7}Li/H $\approx 10^{-10}$, and (ii) the convection processes which are strong enough to carry the superficial Li in the central regions of disk stars, where it can be destroyed, (see, eg. the noticeable low Li abundance in G disk stars where the convection is efficient) do not seem to operate as effectively inside halo stars with similar effective temperatures.

Although there still remains some possibility that lithium could have been partly destroyed in halo stars since their formation 15 to 20 x 10^9 years ago, I will discuss here the implications of a primordial ^{7}Li/H abundance of 10^{-10}.

In conclusion, the implications of the following set of abundances on the primordial phase of the Universe will be discussed (keeping in mind that the quoted values are still controversial and might undergo important changes in the very near future).

$$X_p(D) = 5 \ 10^{-5} \qquad X_p(He^3) = 3 \ 10^{-5}$$
$$X_p(^4He) = Y = 0.23 \pm 0.01 \qquad X_p(^7Li) = 5 \ 10^{-10} \ \text{(by mass)}.$$

2 - 1. 2 <u>Big Bang, particle physics and primordial nucleosynthesis.</u>

The nucleosynthesis processes occurring at the end of the primordial phase of the Universe have been thoroughly explored and reviewed by Wagoner [10, 11] in the frame of the standard or canonical Big Bang models. An important hypothesis which is made in these models is that the Universe is asymmetric, namely, that the baryonic number of the Universe is strictly positive. In other words the amount of antimatter present in the Universe is negligible compared to the amount of matter. This hypothesis is supported by the fact that the ratio between the baryon density n_B and the photon density n_γ is $\approx 10^{-10\pm1}$ and is much higher than it would be in a symmetric universe. ($n_B/n_\gamma \lesssim 10^{-13}$).

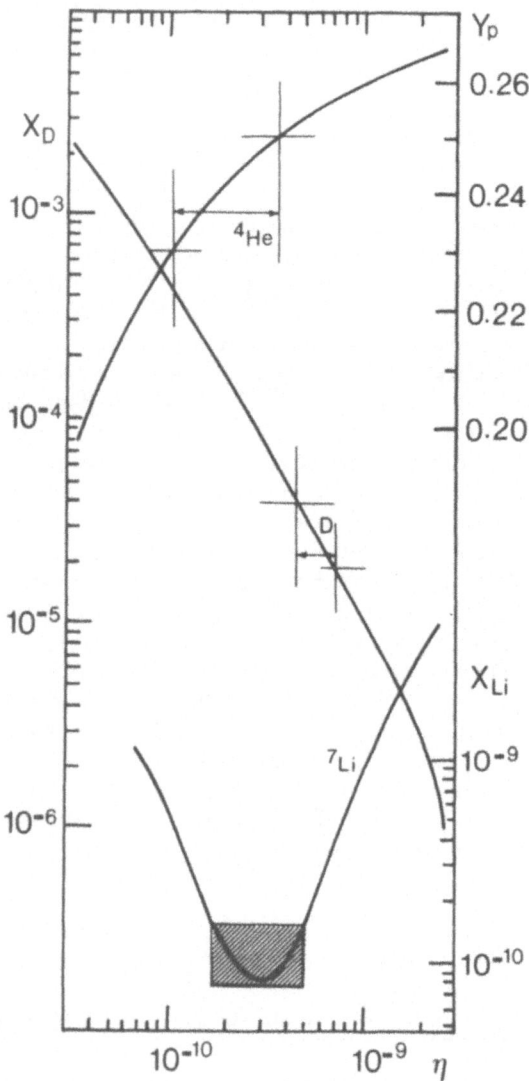

Figure 1.   D, ^{4}He and ^{7}Li abundances plotted against the ratio
between the number of baryons (n_B) and photons (n_γ) $\eta = n_B/n_\gamma$. An
agreement seems to be possible if these abundances reach their
upper limit. This agreement would correspond to a value of $\eta \approx 6$
to 7 10^{-10} i.e. a value for cosmological parameter Ω of $\Omega \approx 0.12$
(adapted from Spite and Spite [7]).

In addition it is assumed that the Universe is homogeneous and isotropic and that its expansion is governed by the general relativity theory ; this is supported by various observations. Finally the hypothesis has been made that there are no new unknown families of leptons (and their related neutrinos). At present, there are three observed families of leptons (the electron, the muon and the tau).

With Big Bang models such as these, D, ^{3}He, ^{4}He and ^{7}Li can be easily explained by nuclear processes occurring at $T \approx 10^9$ K and starting with the $n + p \rightarrow D + \gamma$ reaction. They take place at a time of a few hundred seconds after the birth of the Universe when neutrons and protons start to decouple.

As shown by Wagoner [10] and subsequent studies the D, ^{3}He and ^{7}Li theoretical abundances agree only with the observations for values of the cosmological parameter $\Omega \approx 0.05 - 0.10$, corresponding to the present baryonic densities of the Universe of a few 10^{-31} g cm^{-3}. It has been realized recently that ^{4}He can also be used to select the present baryonic density of the Universe and therefore the cosmological parameter (figure 1).

These values of the cosmological parameter deduced from the formation of the light elements are significantly lower than those deduced from the dynamics of large clusters of galaxies ($\Omega \approx 0.2 - 0.5$).

In order to reconcile the low Ω values deduced from the primordial nucleosynthesis with higher values coming from the large scale structure of the Universe, the following has been proposed : if neutrinos have a mass of about 10 - 30 eV, they should be clumped in clusters of galaxies and provide more mass than the baryons. This hypothesis is not yet supported either by particle physics experiments or by any astrophysical observation, such as the search of a UV emission in the 1000 Å range and in the direction of large clusters of galaxies coming from the decay of massive neutrinos [14]. In any case all values of Ω proposed so far are in favour of an open universe which should expand and cool for ever.

Much work ([12, 13, and 6]) has been devoted to relating the nucleosynthesis to the maximum possible number of neutrino types, to the baryon, photon density ratio, to the neutron life time... The production of helium is related to the neutron/proton ratio, at the time when the neutrons start to decay and are no longer in equilibrium with protons. This ratio is directly related to the expansion rate of the Universe, which increases with the total density of the Universe (including baryons and neutrinos which are almost as numerous as photons and which carry a significant fraction of the overall energy of the Universe). As shown in figure 2, the primordial He abundance adopted here is consistent with a maximum number of three different families of neutrinos (and leptons).

In conclusion the present observations of D, ^{3}He, ^{4}He and ^{7}Li indicate a low baryonic density implying a continuous expan-

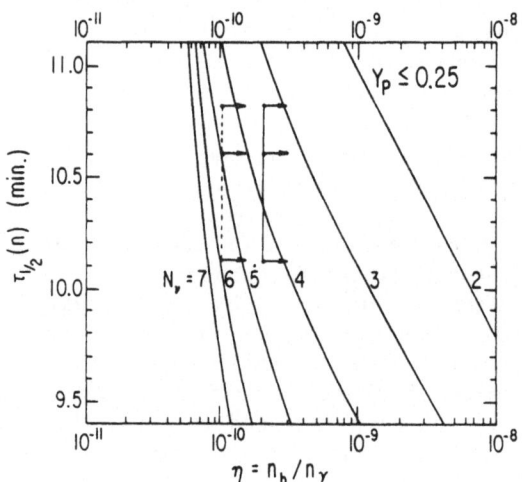

Figure 2.   Sketch drawn by Olive et al. [13] showing the maximum
number of neutrino families for $Y_p \lesssim 0.25$ and with respect to η
(the ratio between the number of baryons and photons) and the
half life of the neutron. The two sets of arrows correspond to
$\eta > 2 \ 10^{-10}$ (dynamics of binary galaxies and small groups of
galaxies, $\Omega \approx 0.04$, solid lines) and $\eta > 10^{-10}$ (dashed lines,
lower limit of the baryon density compatible with the D and ^{3}He
nucleosynthesis).

sion of the Universe. The ^{4}He abundance restricts the number of
neutrino types to about 3, i.e. the presently observed number.
Therefore there exists a strong relation between particle physics
and cosmology : the Grand Unification theories may find their
strongest support from cosmology. Moreover the Big Bang theories
can be challenged by such observations. The light element nucleosynthesis is indeed one of the best cosmological tools presently
available.
        With the new value of the primordial ^{7}Li value Li/H $\approx 10^{-10}$,
^{7}Li ought to be produced during the galactic history. This problem is analyzed in the following sections.

2 - 2 The nucleosynthesis induced by cosmic rays.
        The elements with atomic mass $6 \leqslant A \leqslant 11$ (with the exception
of ^{7}Li which will be treated separately) cannot be produced in
the central regions of the stars because they are easily destroyed by thermonuclear (p,α) reactions. It has been understood since 1970 through the work of Reeves, Fowler and Hoyle [15] that

the bombardment of the interstellar medium by galactic cosmic rays could be the process by which lithium, beryllium and boron are synthetized on the galactic scale. This bombardment induces spallation reactions between the CR particles and the nuclei of the interstellar atoms. In favour of this hypothesis is the fact that one observed a LiBeB/CNO abundance ratio of about 0.2 in the galactic cosmic rays, to be compared with 10^{-5} in the so-called standard abundances.

Meneguzzi, Audouze and Reeves [16] have shown quantitatively by using the "leaky box " model that lithium 6, beryllium and boron can be synthetized in the interstellar medium by the observed cosmic ray GCR fluxes in amounts consistent with their standard abundances. In the "leaky box" model, one assumes that in the galactic disk the GCR sources, which are generally assumed to be the supernovae, are uniformly distributed. In any volume unit of the interstellar medium the number of GCR particles is in equilibrium : GCR particles are produced either at the source or by spallation reactions suffered by heavier nuclei ; they disappear by escaping out of the galactic disk through spallation reactions (interaction with the interstellar medium nuclei) or by slowing down (interaction with the interstellar electrons).

In working out the set of differential equations which take into account all these processes, one finds that, on the average, the GCR particles encounter 8 g cm^{-2} of interstellar matter. This, together with the fact that the ^{10}Be GCR nuclei have had enough time to decay [17], provides a lower limit value for the time elapsed between the acceleration of the CR particles and their entry into the solar cavity of about 3 10^7 years. This corresponds to an average density of the interstellar regions traversed by the cosmic rays $\leqslant 0.1$ particle cm^{-3}.

70 % of the interstellar Li Be B nuclei are produced by interaction between the GCR H and He with the interstellar CNO nuclei. The remaining 30 % come from the slowing down of the GCR Li Be B nuclei. Table 1 provides the results of the Meneguzzi et al. [16] computations. One can observe that the resulting ^{7}Li/^{6}Li ratio is 1.7 instead of 12 while ^{11}B/^{10}B = 2.3 instead of 4.

As noticed first by Meneguzzi et al. [16] and discussed again by Reeves and Meyer [18], a contribution of low energy cosmic rays, the flux of which is strongly affected by the solar modulation, is necessary to increase the ^{11}B/^{10}B ratio : ^{11}B can be produced significantly by the ^{14}N(p,α) ^{11}C(β^+)^{11}B which has a very low threshold. Reeves and Meyer [18] argue that the mixture of 27 % of a low energy CR flux in E^{-3} or 4 % of a low energy CR flux in E^{-5} leads to ^{7}Li/^{6}NLi $\approx$ 6 and ^{11}B/^{10}B $\approx$ 4.

In conclusion, the GCR bombardment of the interstellar medium accounts satisfactorily for the formation of ^{6}Li, ^{9}Be, ^{10}B and ^{11}B but not for all of the ^{7}Li which is observed in young stars and in the interstellar medium. We will see later in section 4.1 - that novae and/or red giant stars are likely sites for the formation of ^{7}Li during the history of the galaxy.

TABLE 1

Formation rate of light elements by the bombardment of the interstellar matter by the galactic cosmic rays. This mechanism accounts rather well for the abundances of ^{6}Li, ^{9}Be, ^{10}B, ^{11}B : column 3 is obtained by integrating the GCR product rate over 10^{10} years and assuming a GCR flux constant during this period

light elements	GCR products rate	Abundance of GCR (relative to H)
^{6}Li	$1.1 \ 10^{-4}$	$8 \ 10^{-11}$
^{7}Li	$1.7 \ 10^{-4}$	$1.2 \ 10^{-10}$
^{9}Be	$2.8 \ 10^{-5}$	$2 \ 10^{-11}$
^{10}B	$1.2 \ 10^{-4}$	$8.7 \ 10^{-11}$
^{11}B	$2.8 \ 10^{-4}$	$2 \ 10^{-10}$

TABLE 2

Physical conditions and by-products of the stellar nucleosynthesis both non explosive and explosive.

	nucleosynthesis during the normal course of the stellar evolution	explosive nucleosynthesis
Hydrogen burning	$T \approx 10^7$ K, $\rho \approx 100 \text{gcm}^{-3}$ by-products: ^{4}He, ^{14}N	$T \ 10^8$ K, $\rho > 100 \text{gcm}^{-3}$ by-products : ^{13}C ^{15}N ^{17}O, ^{11}Ne, ^{22}Ne ...
Helium burning	$T \approx 10^8$ K, $\rho \approx 10^4 \text{gcm}^{-3}$ by-products ^{12}C, ^{16}O, ^{18}O, ^{20}Ne, ^{22}Ne s process	$T \approx 6 \ 10^8$ K, $\rho > 10^4 \text{gcm}^{-3}$ by-products ^{15}N, ^{18}O ^{19}F, ^{26}Al
Carbon burning	$T \approx 6 \ 10^8$ K, $\rho \approx 10^5 \text{gcm}^{-3}$ by-products ^{20}Ne, ^{23}Ne, ^{24}Mg	$T \approx 2 \ 10^9$ K, $\rho \approx 10^5 \text{gcm}^{-3}$ by-products Ne, Na, Mg, Al isotopes
Oxygen burning	$T \approx 10^9$ K, $\rho \approx 10^6 \text{gcm}^{-3}$ by-products ^{28}Si, ^{31}P, ^{32}S	$T \ 3 \ 10^9$ K, $\rho \approx 10^6 \text{gcm}^{-3}$ by-products Si, S, Cl, Ar, K, Co isotopes
Silicon burning	$T \approx 3 \ 10^9$ K, $\rho \approx 10^7 \text{gcm}^{-3}$ Si, S, Cl, Ar, K isotopes	$T \ 5 \ 10^9$ K, $\rho \approx 10^7 \text{gcm}^{-3}$ Ti, V, Cr, Mn, Fe, Co, Ni

2 - 3 The stellar nucleosynthesis.

Stars are responsible for the formation of the nuclei of atomic mass A $\geqslant$ 12. Although the majority of them spend most of their life time on the Main Sequence by transforming H into He in their central regions, the stellar production of He is not sufficient to account for the observed He abundance.

The stellar nucleosynthesis occurs according to two different modes : either during the normal course of the stellar evolution or during explosive phases such as nova or supernova outbursts. Let us consider in turn some aspects of these two modes :

2 - 3. 1 Quiet stellar nucleosynthesis.

An abundant literature [1, 19, 20, 21, 22] has been devoted to present the nuclear and the astrophysical aspects of the processes taking place during the stellar evolution. These are the hydrogen burning occurring in main sequence stars, the helium, carbon and oxygen burning and the slow neutron absorption process (s process) occurring during the red giant phase.

Table 2 summarizes the main characteristics of the nucleosynthetic processes which take place when stars evolve quietly. Due to the restricted format of this review, there is no room here to analyze in detail all these phases. However I would like to make the following remarks :

(i) In current models of chemical evolution of galaxies one distinguishes between elements which can be directly formed (at least in principle) in first generation stars with no metallicity. (These elements are called primary elements) and elements which can only come from the transformation of heavy elements ; these are the secondary elements. Elements such as ^{12}C, ^{16}O, ^{28}Si and ^{56}Fe are primary while ^{13}C, ^{14}N (at least in part), ^{22}Ne, and the s process elements are secondary. It is worth noting that the nucleosynthetic processes leading to the formation of secondary elements are more likely to occur during the normal course of the stellar evolution, while the primary elements are mainly released in supernova outbursts.

(ii) In the central regions of red giant stars where He is transformed into ^{12}C, the triple α ($3\,^4He \rightarrow\,^{12}C$) reaction may often take place in degenerate conditions where the temperature is not controlled by the thermal pressure. In these conditions, the helium flash, which is triggered by this temperature increase induces the partial mixing of the helium and the hydrogen burning zones. Such mixing processes have important nucleosynthetic consequences : for instance, ^7Be can be synthetized by the $^3He + \,^4He \rightarrow \,^7Be + \gamma$ reaction as proposed by Cameron and Fowler [23]. If the convective motions are such that ^7Be can be carried in zones where the temperature is low enough for it not to be destroyed, it can capture an electron and becomes ^7Li. This mechanism could explain the Li overabundance(Li/H $\approx$ 10^{-7}) i.e. 100 times the Solar System value observed in several red giant stars.

The mixing between the H and He burning zone is also at the

origin of the s process nucleosynthesis : the reactions which are
considered as the main neutron sources for such a process ^{13}C
$(\alpha,n)^{16}$O and ^{23}Ne$(\alpha,n)^{25}$Mg do require such mixings.

There are several strong arguments in favour of the occurrence of the s process during the normal course of the stellar
evolution : (i) Technetium, which has radioactive nuclei with
life times $\leqslant 10^6$ years, has been detected in some cool stars ;
(ii) the product of the neutron absorption cross-sections σ by
the abundance N_S of the s process elements has a smooth
behaviour relative to their atomic mass A_S ; (iii) there exist
Ba rich stars and especially the variable star FG Sagittae (see
eg. [24]) which displayed dramatic changes in the abundances of
some s process elements on time scales as short as one year
etc... The study of many explosive processes has been prompted
by the fact that the thermonuclear processes occurring in quiet
stellar interiors are (in general) unable to explain rare nuclear
species like ^{15}N (the observed ^{15}N/^{14}N ratio in the Solar System
is about 30 times higher than the one computed from the CNO
cycle), ^{25}Mg ^{26}Mg ^{29}Si, ^{30}Si......

2 - 3. 2 The explosive nucleosynthesis.

Much work has been devoted to these aspects of nucleosynthesis since the pioneering work of Arnett,Truran, Clayton and
associates in the early 70s. At that time attention was focused
on the explosive C,O and Si burning such as the explosive He
burning (see Table 3 and figure 3 which summarizes the basic
characteristics of these processes and Arnould [25] for a recent
review).

This review is devoted to describing a) some recent progress
which has been made in the explosive hydrogen and helium burning.
These processes are themselves intimately related to the outbursts of the classical novae ; b) recent works where the nucleosynthesis processes triggered by explosions of massive stars have
been thoroughly analyzed and computed in models where dynamical
and the hydrodynamical status of the stellar gas (such as the
radiation transfer and the nuclear reactions) are considered
together. c) A few remarks on the nucleosynthesis of heavy
elements.

a) Nova outbursts and explosive hydrogen and helium burning.

The characteristics of the classical novae are the following : they are white dwarfs of about 0.5 - 1M$_0$ which belong to
a binary system and therefore which accrete hydrogen-helium rich
material coming from their companion. This material falls supersonically on the surface of the prenova and heats the prenova
material which is enriched into C,N,O. This enrichment has been
observed by many authors (see eg. Friedjung, [26] for a review of
these abundance determinations). The nova CNO abundances relative
to H are a few tens larger than in the Solar System. The heating
due to the supersonic accretion triggers the so-called hot CNO
cycle of nuclear reactions (see eg. Starrfield et al., [27])

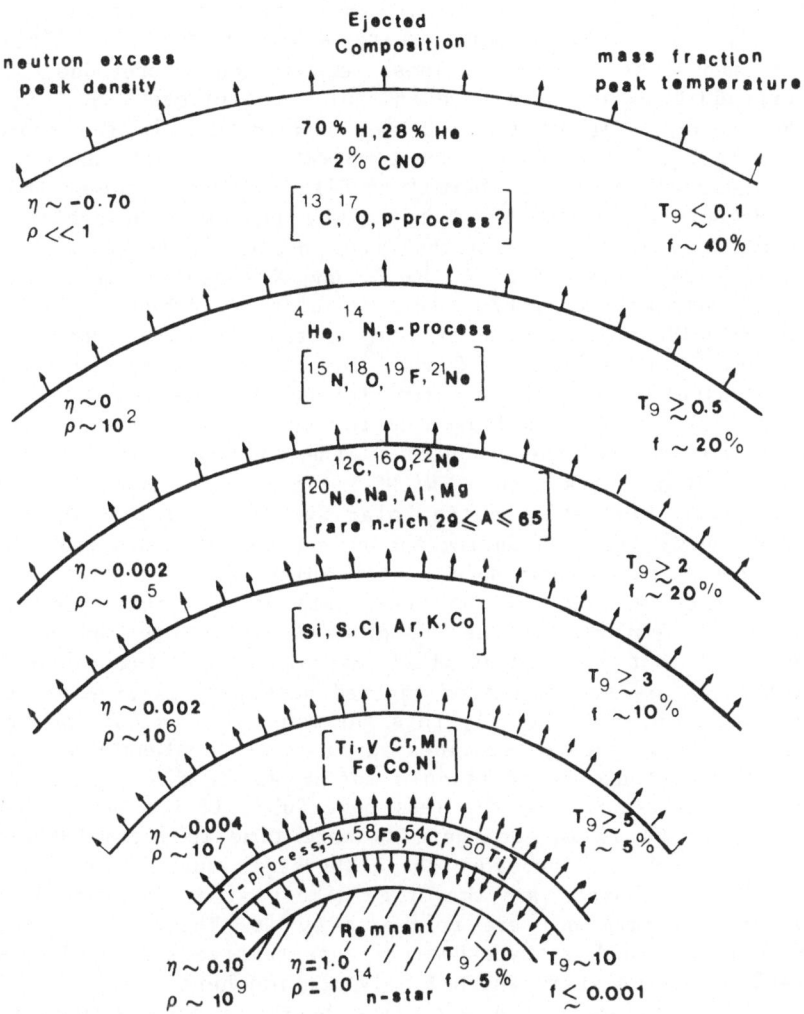

Figure 3. Theoretical onion skin model for a large mass star $(M > 10M_\odot)$ before the supernova explosion. Each layer is characterized by its density ρ (in g cm^{-3}) its temperature T_9 (in units of 10^9 K) its relative mass f, its neutron excess $\eta \approx |N-Z|/N+Z$ and its chemical composition.

which itself makes the outer layers of this object explode. The models of Starrfield et al., [27] based on the occurrence of such reactions (studied also by Audouze et al., [28] and Lazareff et al., [29] account in a straightforward way for the light curve of the nova outbursts, the ejected mass per event (about 10^{-5} to 10^{-3} M_0) and predict some enhancements of the ^{13}C, ^{15}N, ^{17}O, ^{21}Ne and ^{22}Ne elements. These enhancements are due to the overproduction of their beta instable progenitors ^{13}N, ^{15}O, ^{17}F, ^{21}Ne, ^{22}Na, ^{22}Mg if the reaction temperature in the exploding layer is $T > 2 \ 10^8$ K. At such temperatures the beta decays which are independent of the thermal conditions become slower than the thermonuclear reactions. That is why the beta unstable nuclei become more abundant than the other nuclei in these conditions. There is an observation of the C, and N isotopic composition in the CN molecule concerning Nova Herculis performed by Sneden and Lambert [30] which seems to confirm the relative enrichment of ^{13}C and ^{15}N relative to ^{12}C and ^{15}N. This overproduction of these beta unstable nuclei have two interesting consequences : (i) novae might be the nucleosynthetic sources of elements like ^{15}N which are indeed underproduced in the hydrogen burning zones of non exploding stars ; the $^{15}N/CNO$ ratio in such zones is 30 times lower than this ratio in the Solar System. (ii) Most of the nuclear energy released during the nova outburst comes from the beta decay of these unstable nuclei which act like energy reservoirs which account for the time scale of the light curves. From the above features of these explosive cycles one can be easily convinced that the energy which is released during the outburst (and therefore the amount of ejected material) is proportional to the CNO abundance. Finally this thermonuclear mechanism triggering such outbursts provides a natural explanation for the Kukarkin-Parenago correlation : the novae in which outbursts are more frequent are also the less powerful : if the period between two outbursts is larger there is more time to accumulate exploding material and therefore the outburst is more important.

Nova outbursts are therefore likely sites for the explosive hydrogen burning and are triggered by it. There are other possible sites, such as hypothetical supermassive stars [31] and accreting neutron stars which I only mention here.

Until recently the explosive hydrogen burning models were limiting their analysis to the nuclei up to ^{25}Mg. Wallace and Woosley [32] have shown that explosive hydrogen burning can affect heavier chemical elements if the temperature is high enough to transform the CNO elements into more complex nuclei. Figure 4 extracted from their work shows that when the temperature and the density of the exploding material are such that its representative point is above curve (a) ($T \approx 5 \ 10^8$ K $\rho \approx 1g \ cm^{-3}$ for instance), this material experiences the "classical" hot CNO burning evaluated for instance by Audouze et al., [28]. If the representative point (ρ,T) is above curve (b) (example : $T = 8 \ 10^8$ K, $\rho = 100 \ g \ cm^{-3}$) the CNO nuclei are transformed into heavier elements

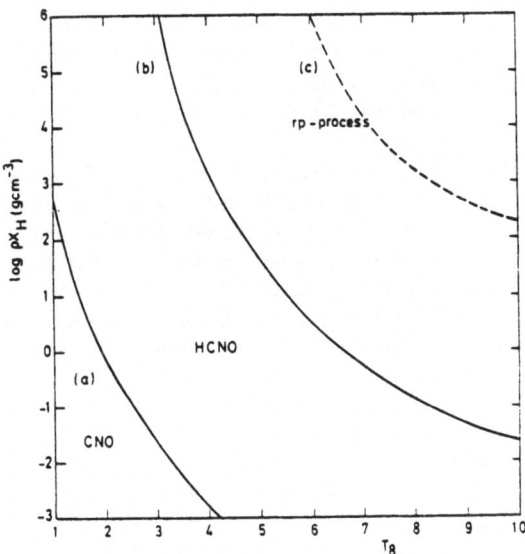

Figure 4.   The different modes of hydrogen burning in the temperature-density plane. Below curve (a) : hydrogen burning proceeds non explosively through the "cold" CNO cycle). Between curves (a) and (b), the hydrogen burning is explosive ("hot" CNO cycle). Between curves (b) and (c) the CNO nuclei are transformed into heavier ones by the $^{15}O(\alpha,\gamma)$ $^{19}Ne(p,\gamma)$ ^{20}Na chain. Above curve (c) the (α,p) reactions play a significant role in the so-called rapid p process from Wallace and Woosley [32].

like Ne, Mg and so forth because the $^{15}O(\alpha,\gamma)^{19}Ne(p,\gamma)^{20}Na$ chain of reactions occurs very rapidly (remember that most of the CNO nuclei are transformed into ^{15}O by the hot CNO cycle). According to Wallace and Woosley [32] the explosive hydrogen burning can proceed through the hot NeNaMg cycle up to the iron-peak nuclei. At the highest temperature and density (T $\approx$ 10^9 K, ρ $\approx$ 10^4 g cm^{-3}), the (α,p) reactions play a major role in tranforming the lighter nuclei into heavier ones. The reader is referred to Wallace and Woosley [32] or Arnould [25] for more details. In any case there are so many uncertain reaction rates in these nucleosynthetic chains that the outcome of the explosive hydrogen burning is still debatable.

       To end my remarks on this nucleosynthetic process, there are two interesting by-products which deserve a special mention ^{22}Ne and ^{26}Al. As shown by various authors, Vangioni-Flam et al. [33], Audouze et al. [34], Arnould [35], Arnould et Norgaard, [36],

the explosive hydrogen burning might produce ^{22}Ne which exists in
large amounts in the Ne E gaseous phases of some carbonaceous
chondrites like Orgueil, and in the galactic cosmic rays.In this
last case we have advocated that the determination of the Ne
isotopic composition in the cosmic rays might provide some clue
to their origin, namely that cosmic rays are accelerated in the
vicinity of exploding objects where about 80 % of interstellar
material has been mixed with 20 % of matter processed by the
explosive hydrogen burning.

Concerning ^{26}Al which is detected as ^{26}Mg in some Al rich
silicate phases of Allende and which sets a limit of a few mil-
lion years between the end of the nucleosynthesis and the isolation of the protosolar nebula, the explosive hydrogen burning
might explain its formation, as shown by Arnould et al.[36]. It
should be mentioned that this time limit for the explosive H burning is about 100 times lower than the time limit deduced from
some short lived r process elements like ^{129}I or ^{244}Pu. This
would mean that the r process occurs only in a small fraction
($\approx$ 1 %) of all the exploding objects.

The explosive helium burning which takes place at T $\approx$ 7 - 8
10^8 K and for $\rho \approx 10^4$ g cm^{-3} should be a more effective way to
produce ^{26}Al (together with some ^{15}N, ^{18}O and ^{19}F) while this
process is unable to produce large amounts of ^{22}Ne (Arnould and
Norgaard [37], Vangioni-Flam et al. [33]).

b) Some recent models of explosive stellar nucleosynthesis.

Several groups including the Livermore-Santa Cruz team (T.
Weaver, S. Woosley and associates), W.D. Arnett, W. Hillebrandt,
F.K. Thielemann... have attempted to treat together the stellar
evolution, the hydrodynamics of the stellar explosions and the
nuclear processes occurring during the explosions. There are indeed many ways of analyzing the outcome of some processes of explosive nucleosynthesis. The abundance N_a of a given element a
is given by :

$$\frac{dN_a}{dt} = - \sum_b N_a N_b <\sigma V>_{ab} + \sum_{c,d} N_c N_d <\sigma V>_{c,d}$$

where the first term represents the destruction of a by all the
a(b,x)y reactions and the second the formation of a by all the
c(d,z)a reactions.

The first approximation is to assume that the abundance of a
has reached an equilibrium and to put $dN_a/dt = 0$ in the above
equation. The second approximation is to assume that the reactions take place at fixed densities ρ and temperatures T, which
means that the reaction rates $<\sigma V>$ are constant. The third
approximation considers a parametric variation of ρ and T such as
the one deduced from the free-fall adiabatic approximation where
$\rho = \rho_0 \exp(-t/tc)$ and $T = T_c \exp(-t/3t_c)$ and $t_c \approx 500/$
$\sqrt{\rho_0(g.cm^{-3})}$ sec. The fourth approximation uses the T and ρ variations deduced from a stellar model where the variation of the
energy released by the nuclear processes is crudely evaluated.

Now one tries to solve the full problem (evolution and nucleosynthesis) with a minimum number of assumptions. The models constructed by Weaver and Woosley [38] concern stars of 15$M_\odot$ and 25$M_\odot$. They first run classical stellar models in which the functions dP/dr, dM/dr, dL/dr, dY/dr giving the pressure, the mass, the luminosity and the temperature in any stellar region together with the equation of state f(P,ρT) and the opacity are computed throughout the life time of the star. In their equation of state they take into account the electron (and positron) degeneracy, while the nuclei are non degenerate and are in thermodynamical equilibrium. The convection motions which govern the energy transport are estimated with the mixing length formalism. They treat as accurately as possible the nuclear physics aspect of the model with equations like the one quoted above. The hydrodynamical treatment of the explosion is made in a Lagrangian, implicit framework where one introduces some artificial viscosity to avoid any strong discontinuities in densities and temperatures.

The results of the outcome of the explosion of a 15$M_\odot$ and a 25$M_\odot$ star are displayed on figures 5 and 6. One can notice from figure 6a that a 15$M_\odot$ star explosion produces a satisfactory amount of elements from C to <Fe> : the model provides a better account of heavier elements Mg to Fe than of lighter ones. The outcome of a 25$M_\odot$ star explosion might be a little bit better.

Although there is a significant improvement in the nucleosynthetic analysis of an explosive process one should realize that many uncertain simplifications are still made : rough stellar models with a crude treatment of the convective motions ; uncertain nuclear reaction rates, approximate hydrodynamical treatment due to the unvoidable introduction of the artificial viscosity. Improvements on almost all aspects of these quite complex computations are needed but are extremely hard to achieve for two reasons (i) too many different fields of physics are involved in these models ; (ii) they require formidable computing facilities and time. Nevertheless, it is to this type of problem that nuclear astrophysicists should now devote their efforts.

c) A few words concerning the neutron absorption processes.

The arguments in favour of the slow neutron absorption (s process) have been presented in section 2 - 3. 1. The best proof of the occurrence of rapid neutron absorption (r process) is the existence of the heavier stable isotopes which cannot be synthetized by the s process nucleosynthesis such as the existence of translead nuclei because the s process stops at the synthesis of ^{209}Pb. Table 3 presents the list of problems which still have to be solved in order to fully understand these neutron absorption processes. Concerning the s process, there is still some ambiguity about which reactions are actually the most efficient neutron sources. There is still a debate between the choice of the ^{13}C(α,n) ^{16}O reaction or that of the ^{22}Ne(α,n) ^{25}Mg reactions.

(a)

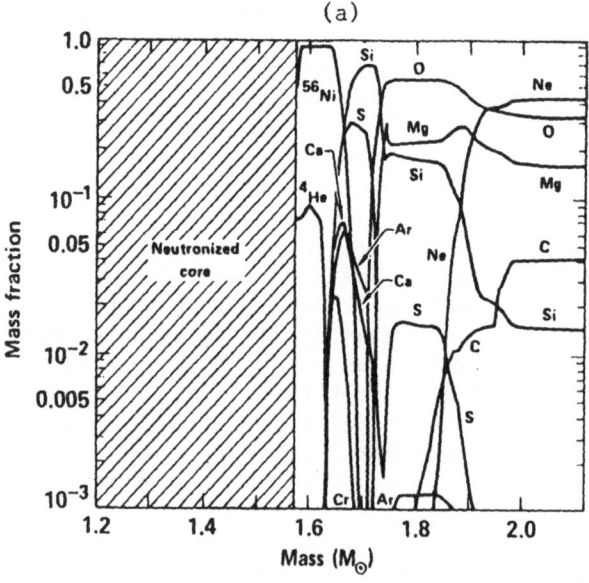

(b)

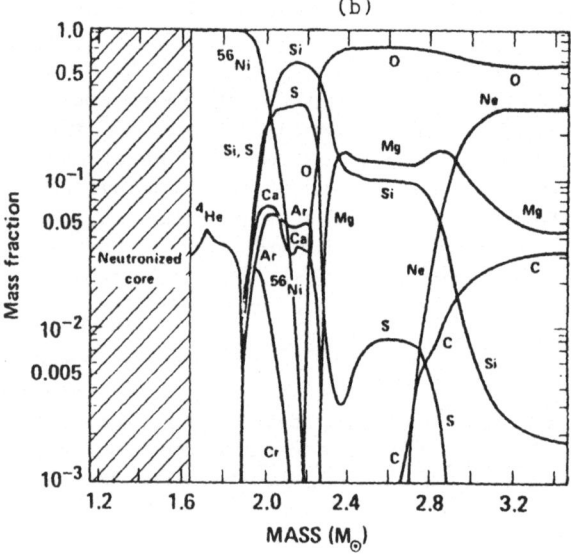

Figure 5. Abundances of some major species in various processed
layers of 15 M$_\odot$ and 25 M$_\odot$ stars (respectively figures 5a and
5b). The cores contain large amounts of neutron rich species.
Outside the cores, the abundances result from the explosive
burning of Ne,O and Si. No significant C burning takes place in
these layers while the more external layers are not affected by
the explosion (from Weaver and Woosley [38]).

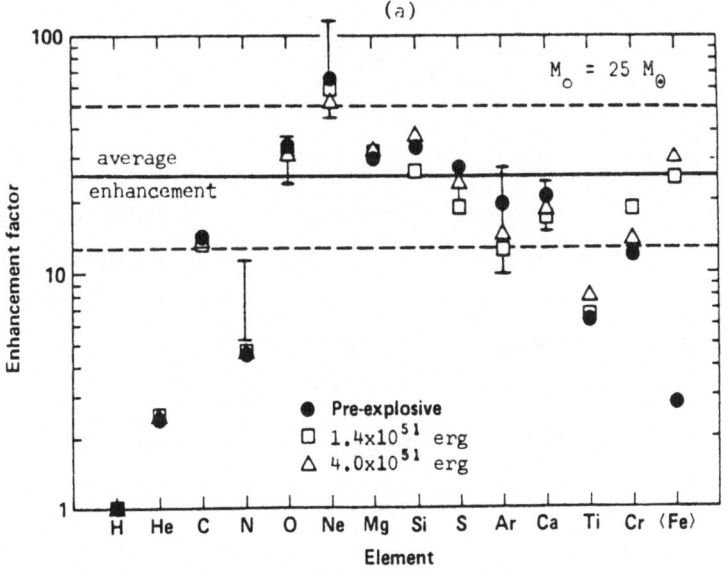

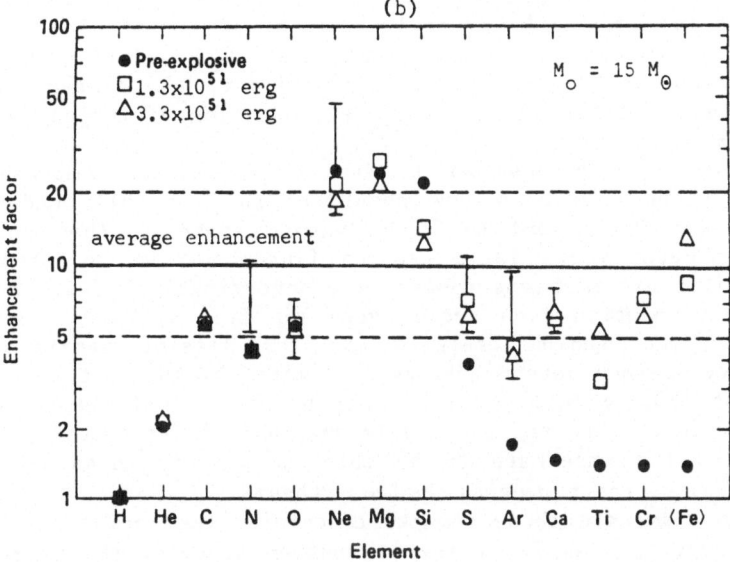

Figure 6. Enhancement factors or different elements up to the Fe peak (represented by <Fe>) for stars of 15M$_\odot$ (figure 6a) and 25M$_\odot$ (figure 6b). These factors are defined as the mass of an element outside the neutronized core relative to the hydrogen mass and normalized to the corresponding Solar System abundance (From Weaver and Woosley [32]).

TABLE 3

	s process	r process
Astrophysics problems	neutron sources	thermodynamics (T,ρ)
	convection pulses	and hydrodynamics of
	dredge up mechanisms	the explosive events
Nuclear physics problems	absorption cross-sections	nuclear masses
	isomeric states	binding energies
		fission barriers

=====

Moreover the convective motions, pulsations and/or dredge up mechanisms which can mix the hydrogen and the helium burning zones in red giant stars and therefore trigger the neutron production, have to be properly taken into account. A systematic determination of the neutron absorption cross-sections should be undertaken, for example the determination of the characteristics of isomeric states influencing the production of nuclear species such as ^{176}Lu.

For the r process which leads to abundances comparable to those of the s process, one should be able to fully understand the thermodynamic and the hydrodynamic state of the exploding stellar gases where large neutron fluxes can be released. The most difficult problems concern the determination of the nuclear masses of neutron rich nuclei very far from stability, such as the neutron binding energies and the fission barriers. The accuracy of such determinations is limited by the fact that we do not yet have a full understanding of the actual nature of the nuclear force. The nuclear models available in present literature are much too uncertain to be able to provide reasonably good estimates of the r process nucleosynthesis.

To terminate these remarks concerning the formation of elements heavier than Fe, there is another process, the p process, which should explain the formation of rare nuclei having a relatively large number of protons compared to their neutron numbers. Although I proposed some time ago, in collaboration with J.W. Truran, some mechanisms to explain the p process ([39] see also [40]) by arguing that this process should occur in explosive hydrogen burning zones, this still poorly known nucleosynthetic process has to be studied again.

2 - 4 Summary.

To summarize we have identified three possible sites for nucleosynthesis (i) the primordial nucleosynthesis is responsible for the formation of the lightest nuclear species D, ^{3}He, ^{4}He, ^{7}Li. The nucleosynthesis of these elements seems to indicate that the baryonic density of the Universe is $\approx$ 3 to 5 10^{-31} g cm^{-3}. It seems to imply an open Universe expanding for ever. The ^{4}He primordial abundance is consistent with a number of different types of neutrinos equal to the presently observed number of 3. Forthcoming work might question these simple conclusions : the agreement between the present density of the Universe deduced from ^{4}He, D, or ^{7}Li for instance is not as good as it should be to confirm that Standard Big Bang models are those which explain this nucleosynthesis satisfactorily.

(ii) The nucleosynthesis at the galactic scale, induced by the cosmic ray bombardment of the interstellar medium well explains the formation of the ^{6}Li, ^{9}Be, ^{10}B and ^{11}B. One should be aware however, that one has to call on the effect of unseen low energy cosmic rays to account for the observed ^{11}B/^{10}B ratio.

(iii) The stellar nucleosynthesis is responsible for the formation of most of the observed nuclear species. The overall nucleosynthesis scheme is now fairly well understood. All the difficulties presented by such processes are not yet fully solved : there is still the solar neutrino problem which casts doubt on the current stellar models and/or for some cross-section determinations related to the hydrogen burning cycles. Explosive objects like novae and especially supernovae are indeed the most powerful nucleosynthetic "factories". Although much effort is now made to account for the hydrodynamics of such explosions, the full physical account of such explosions is far from being completely and satisfactorily covered. Finally processes like the p process, responsible for the rare heavy nuclei with large proton numbers, and the r (rapid neutron absorption) process, are not properly understood today.

There are many interesting studies yet to be done on the primordial nucleosynthesis before all the discrepancies mentioned above can be solved. The outcome of supernova and nova explosions must be analysed in a convincing way in order to solve the questions still posed by the r and the p process. These are some of the most difficult puzzles of future studies in nuclear astrophysics.

3 - The chemical evolution of galaxies

Although there are still many unsolved problems which prevent a full understanding of all the nucleosynthetic processes responsible for the formation of the observed chemical species, the general scheme (i.e. the relative role of the primordial nucleosynthesis, the galactic cosmic rays and the stars of any mass) should not vary too much. One is then tempted to analyse why significant abundance variations are observed from one old

star to a young one or from one galactic region to another. For
instance the old halo stars are known to be very deficient in
heavy elements compared to the average disk stars. At the surface
of these stars the Fe abundance can be as low as 0.1 to 1 % of
the Solar System value and one notices that internal regions of
the disk of our own Galaxy and nearby spirals (M 31, M 33, M 101
for   instance)   are   more   metal   rich   than   external   regions.
The composition of matter does evolve during the history of the
Galaxy. In this review I would like to provide an account of the
principles guiding the analysis of the chemical evolution of galaxies, for example the recent studies in which we have been involved among them stochastic effects on the chemical evolution,
the evolution of ^{7}Li, the rate of star formation deduced from
planetary nebulae distributions etc... Some references concerning
chemical evolution of galaxies are : Audouze and Tinsley [41],
Tinsley, [42], Pagel and Patchett [43], Guelin and Lequeux [44],
Alloin et al. [45], Wannier [46] and Pagel and Edmunds [47].
There are at least two other ways by which a galaxy can evolve :
like any other material object, a galaxy evolves dynamically
especially during its formation. Moreover, since a galaxy is
basically a collection of stars, the luminosity and the colors of
which vary with time, the overall luminosity and spectral types
evolve as well. This review is not concerned with these two
aspects, which nevertheless cannot be entirely ignored.

3 - 1 The relevant observations.
      The current models of galactic evolution have to take into
account four groups of observations : the overall characteristics
(total mass, total luminosity) of the galaxies ; the distribution
of the interstellar gas ; the distribution of stars ; and the
variations of the element abundances with time such as those of
some well analyzed isotopic ratios.
      1 - The galaxies are still classified according to the taxo-
nomy proposed by Hubble (modified by some authors like Van den
Bergh). One distinguishes three classes of galaxies. a) The
elliptical galaxies are characterized by a large mass/luminosity
M/L ratio (M/L $\approx$ 50 in solar units), the color of these galaxies
is mainly red which means that they are dominated by low mass
stars. The content of interstellar gas is insignificant in comparison with that of the stellar mass (m_g/m_{tot} < 1 % and their
metallicity is high Z $\gtrsim Z_0$). b) The spiral galaxies (our own
Galaxy is a member of this class) are characterized by the fact
that they have two major components : a central component, often
called the "bulb", which looks like a small elliptical galaxy
(with a lower M/L ratio, however) and the disk in which one notices the presence of spiral arms. There are noticeable color and
abundance gradients : the heavy element abundances are bigger (by
factors of about 10) in the central regions, compared to the
external ones which are blue while the bulb is yellow/red. The
interstellar gas density varies also from less than 1 % in the

central regions of a spiral up to about 10 % in the external regions. Nevertheless, all throughout the galactic disk of spirals
the M/L ratio is about 5 to 10 (in solar units) which means that
the spirals in average are more luminous and less massive than
the ellipticals. c) The irregular galaxies are bluer and have a
lower M/L ratio (of a few solar units) a larger interstellar gas
density (up to 15 %) and a lower metallicity ($Z < Z_0$).

The differences between these three morphological types of
galaxies are generally interpreted as a consequence of their mass
difference. The galaxies are assumed to have about the same age :
they should have been formed just at the end of the radiative era
i.e. 10^6 to 10^9 years after the Big Bang. The more massive galaxies have evolved very rapidly, i.e they have formed stars on a
very short time scale. These stars have "swallowed" most of the
interstellar gas and have rapidly enriched the galactic matter
into heavy elements. By contrast, the less massive galaxies form
stars on a longer time scale, which means that some interstellar
gas is still left to form massive stars, and that the heavy element contamination has not been as effective as in the massive
galaxies (Table 4).

2 - Regarding the distribution of the interstellar gas and
the stars, it has been observed that galactic regions in which
the interstellar gas density is higher have,on the average, a relatively higher density of high mass stars. In most of the current models of galactic evolution, one assumes that the rate of
star formation is related to the interstellar gas density in a
form such as that proposed by Schmidt $dS/dt = \mu^n$ where S is the
stellar mass fraction, μ the interstellar gas fraction (both normalized to 1 such that $S + \mu = 1$) and n an exponent which is ge-

TABLE 4

Galaxy	m_g/m_{tot}	$(O/H)/(O/H)_0$	$\tau_* Mg)/(\tau_* Mg)0$
Solar Neighborhood	0.06	1	1
NGC 6822	0.11	0.43	0.28
LMC	0.12	0.56	0.25
NGC 4449	0.15	0.60	0.15
IC 1613	0.21	-	0.17
IC 10	0.24	0.36	-
IIZw70	0.29	0.26	0.54
IIZw40	0.34	0.28	0.20
SMC	0.42	0.17	0.12
IZw18	?	0.03	0.19

From Lequeux [47]

nerally taken as $1 < n < 2$.

3 - It has been stated at the beginning of this section that the old halo stars are very metal deficient in comparison with the population I disk stars. It should be kept in mind that the number of metal deficient stars is much lower than that of stars with solar abundances, and that no star has been observed, so far, with a zero metal abundance.

Concerning the disk stars, recent work performed by Twarog [49] based on photometric surveys of many stars in the Solar Neighborhood shows that there is an enrichment in the Fe abundance of about a factor 4 between the oldest disk stars and the youngest. This enrichment which takes place in times of about 16 10^9 years has to be compared with enrichments of factors 10^3 occurring during the first 10^9 years of galactic life. Gradients in the metal composition of galactic disks of spirals have been observed in nearby galaxies and in ours (Shaver et al., [50] showing significant enrichments in internal regions of these objects.

4 - Isotopic ratios concerning elements like C, N, O and S have been measured in different molecular clouds located at various galactocentric distances. Table 5 gives a summary of the relevant data. One notices in particular an enrichment of ^{13}C versus ^{12}C in the interstellar medium relative to the Solar System and especially noticeable in the Galactic Center, an enrichment of ^{14}N relative to ^{15}N and ^{17}O and ^{18}O relative to ^{16}O. These determinations provide interesting contraints on models of chemical evolution since the Solar System material is $4.6\ 10^9$ years old, while the interstellar medium can be considered as being very recent, since it is being contaminated by the most recent explosions, winds , etc...

3 - 2 Current models of chemical evolution.

The models which can be designed to describe the chemical evolution of galaxies follow in general some basic prescriptions which have been clearly set up e.g. by Tinsley [42]. One should

TABLE 5

Solar System and Interstellar isotopic ratios
(from Wannier, 1980)

Region	$^{12}C/^{13}C$	$^{14}N/^{15}N$	$^{16}O/^{18}O$	$^{18}O/^{17}O$	$^{32}S/^{34}S$
Solar System	89	270	500	5.5	23
ISM (Solar Neighborhood)	60	330	500	3.2	20
ISM (Galactic Center	26	550	300	3.3	18

evaluate at any given time t the total mass of the region considered M, the fraction of interstellar gas μ = Mg/M, the mass of
stars M_s = $(1-\mu)$M and the metal (or any chemical species abundance Z. $Z_g M_g$ is the mass of metal in the gas and $Z_s M_s$ is
the mass of metal in the stars. One needs also to define ϕ(m)dm
the initial mass function of the stars i.e. their relative distribution according to their mass at the time of their birth. The
function ϕ(m) follows generally a $m^{-\alpha}$ dependence referred to as
the Salpeter law. Much work has been devoted to determining the
exponent α based on stellar statistics. In the original Salpeter
law α = 2.35. From the analysis of the Solar Neighborhood i.e.
the region located within 1 kpc from the Sun (it as a torus shape
due to the galactic rotation) one can deducethe exponent α of the
IMF. $\alpha+1$ = -0.25 for $0.4 \leqslant m \leqslant 1.0$; $\alpha+1$ = -1.0 for $1 \leqslant m \leqslant 2$;
$\alpha+1$ = 1.3 for $2 < m < 10$; and $\alpha+1$ = 2.3 for $10 \leqslant m \leqslant 50$ where m
is the mass in solar units. The IMF is flatter in the low mass
range.

The star formation rate is noted ψ(t) ; the total ejection
rate from stars (i.e. the amount of matter which returns from
stars to the interstellar medium is E(t) and the net inflow rate
of interstellar gas (coming from regions external to the one
under study) is f.

Given these definitions one can write the following set of
equations :

$$\frac{dM}{dt} = f$$

$$\frac{dMs}{dt} = \phi - E$$

$$\frac{dMg}{dt} = -\phi + E + f$$

$$E(t) = \int_{m_t}^{\infty} (m-r_m) \ \psi(t-\psi_t)\phi(m)dm$$

In this last equation giving the ejection rate E(t), r_m is the
mass of the remnant left at the end of the evolution of a star of
mass m, and τ_m is its life time and m_t is the mass of the
star for which τ_m = t.

In analytical approximations one often assumes that the
stellar life time τ_m is negligible compared to t (which is a
valuable approximation for the stars massive enough to contribute
to the galactic enrichment in metals).

Concerning the evolution of the metal abundance Z, one can
write :

$$\frac{d(ZMg)}{dt} = -Z\psi + E_Z + Z_f$$

if one assumes that material ejected form stars is mixed in the
interstellar gas at once and that the stellar Z production does

for primary elements defined in section 2 - 3.1 and not for secondary ones).
The first term of the r.h.s is the amount of metal which goes into stars ; the third term is the contribution coming from the external matter. The total ejection rate of metals from stars is :

$$E_Z = \int \left[(m-r_m - m_{pZm}) \, Z(t-tm) + m_{pZm} \right] \, \psi \, (t-\tau_m)\phi(m)dm$$

where pZm is the mass fraction of a star of mass m which is
ejected as pure metal Z (ejected and processed). The quantities
pZm and r_m which are used in our own models are illustrated in
figure 7 sketched by Alloin et al., [45] who followed the current
nucleosynthetic prescriptions of Arnett for high mass stars and
of Iben and Truran for low mass stars.
   E_Z is then the sum of two terms : the first represents the
ejection rate of unprocessed metal, while the second represents
the rate of ejection of metal which has been synthetized by the
stars under study.

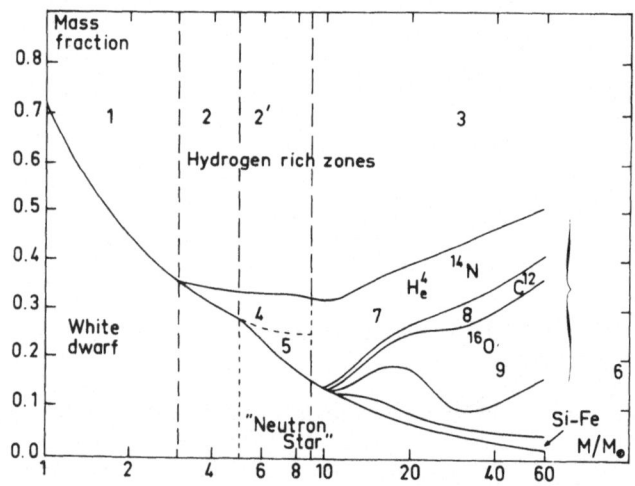

Figure 7. Mass fraction of a star with a given mass ejected into
the interstellar medium or fossilized as white dwarf or neutron
stars. The ejected mass can be enriched into heavy elements for
stars with mass above $4M_\odot$ (from Alloin et al. [45].

   Finally it is useful to define an important function which
is the yield y : the mass of new metals which are ejected by the
stars by unit of stellar mass.

$$y = \frac{1}{1-R} \int_{m_1}^{\infty} m_p Z_m \, \phi(m) dm$$

where m_1 is the present turn off mass and R the returned fraction

$$R = \int_{m_1}^{\infty} (m - r_M) \phi(m) dm$$

the notion of yield is quite important because the asymptotic limit of Z is y for primary elements and $(Z_p \, y)$ for secondary elements.

In principle for all the above quantities one can set up either simple computer codes to follow up the evolution with time of the element abundances and the gas density or to make simple approximations. The standard ones are made in the so-called "simple model" for which at time t = 0, S = 0, μ = 1, Z = 0; where the matter is assumed to be fully mixed, where one makes the instant recycling approximation, where the rate of star formation and the initial mass formation are supposed to be time-invariant and for which the considered region is assumed to be closed without any inflow contribution.

In the frame of the "simple model" one obtains very simple solutions

$$\mu = \exp(-t/\tau_0) \qquad \text{if } dS/dt = \mu \text{ (n=1 in the Schmidt law)}$$

$$\mu = \frac{1}{1+t/\tau_0} \qquad \text{if } dS/dt = \mu^2 \text{ (n=2)}$$

where τ_0 is the characteristic scale of conversion of gas into stars,

$$Z = y \ln(1/\mu)$$

and $\quad \dfrac{S(Z)}{S_1} = \dfrac{1 - \mu_1^{Z/Z_1}}{1 - \mu_1}$

(see derivation in Pagel and Patchett, [43])
where S(Z) is the stellar density corresponding to the metallicity Z. S_1 and Z_1 are the present stellar density and metallicity for the Solar Neighborhood.

It is known that the "simple" model fails (i) to describe the Z(t) dependence : Z(t) varies too steeply after the first 10^9 years compared to the observations (ii) concerning the so-called F-G dwarf problems identified by Pagel and Patchett [43] : the relation $S(Z)/S_1$ predicts too many stars of low metallicity in comparison with the observations.

To alleviate these difficulties one can leave one or more of the hypotheses which are made in the "simple model" and take into account (i) possible infall of external gas which prevents the metallicity from increasing too rapidly with time ; (ii) assume that the initial mass function varies with time namely that the exponent α was on the average smaller in the past, favouring the

formation of massive stars in the early stages of the galactic
evolution ; (iii) assume that there was at the very beginning of
the history of the galaxy a first generation of massive stars,
which very quickly provided the low but non zero metallicity observed in population II stars.

   In studying the evolution of the C and N isotopic ratios
with time (see e.g; Vigroux et al. [51] ; Audouze et al. [52] we
used equations derived from the simple model, but with the possibility of infall (or inflow) of external gas. We avoided the instant recycling approximation which leads to spurious results when
one analyzes central regions of our Galaxy, where the gas mass
density is less than 1 % of the total mass density. The equations
describing the evolution of the gas density and of the element
abundances are :

$$\frac{d\mu}{dt} = -\nu\mu + \int_{m_1}^{m_u} \frac{E(m)}{m} \phi'(m) \, \nu\mu(t-\tau_m)dm + \delta$$

$$\frac{dZ}{dt} = -\nu Z + \int_{m_1}^{m_u} A_Z(m) \, \phi'(m) \, \nu\mu(t-\tau_m)dm + Z_f$$

where ν is related to the characteristic time scale of gas
processing into stars ; $\nu = 0.2$ to 0.3 for Solar Neighborhood
models corresponding to time scales $\tau_0 \approx 5 \ 10^9$ years ; while
τ 2 for central regions of our Galaxy correponding to time
scales $\tau_0 \approx 5 \ 10^8$ years. E_m represents the stellar gas fraction returning to the interstellar medium, and $A_Z(m)$ is the
mass fraction of the element Z released by the stars at the end
of their evolution.

   These formulae have been applied to the evolution of the C
and N isotopes and the reader is referred to Vigroux et al. [50]
and Audouze et al. [51].

   In the case of deuterium the equation giving the evolution
with time of its abundance is very simple because D is entirely
destroyed when it goes into stars ; one has
   $d\sigma_D/dt = \nu\sigma_D + \delta$

the equation of which is (Audouze et al., [53]

$$\frac{D}{H} = \frac{D}{H_{prim}} \left[\frac{1}{\mu} (1 - \frac{\delta}{\nu} \exp(-\nu t) + \frac{\delta}{\nu} \right]$$

   In the Solar Neighborhood with the characteristic stellar
processing time scales deduced from other isotopic evolution the
ratio
   $0.3 \leq (D/H)_{present} / (D/H)_{primordial} \leq 0.5$

means that on the average about 50 % of the primordial deuterium
has been destroyed during the galactic evolution ; this number

might be less if there is a significant infall of external gas having the primordial composition.

In the Galactic Center the D abundance should be lower than in the solar neighborhood, since the CH_3D/CH_4 ratio observed in molecular clouds like Sgr A and Sgr B_2 is ten times lower than the same ratio observed in molecular clouds more distant from the center. But if one takes a model where $\nu \approx 2-3$ with no infall, the resulting D/H) is as low as 10^{-12} (compared to D/H $\approx 10^{-5}$ in the Solar Neighborhood) which does not seem consistent with a $CH_3D/CH \approx 10^{-4}$. This means that in the Galactic Center either there is enough infall of external gas with primordial abundances, or the galactic cosmic rays produced in such regions have somewhat larger fluxes than those observed in the Solar Neighborhood. In this case some deuterium could be produced by spallation reactions triggered by the galactic cosmic rays on the central interstellar medium. (Audouze et al. [53].

4 - A few recent works on chemical evolution of galaxies.

4 - 1 The evolution of lithium 7.

In this section, I refer back to the ^{7}Li abundances deduced from the work of Spite and Spite, [7]. According to these authors ^{7}Li/H primordial $= 10^{-10}$ while ^{7}Li/H at the formation of the Solar System $= 10^{-9}$ and ^{7}Li/H$_{present} > 5 \cdot 10^{-10}$ (figure 8).

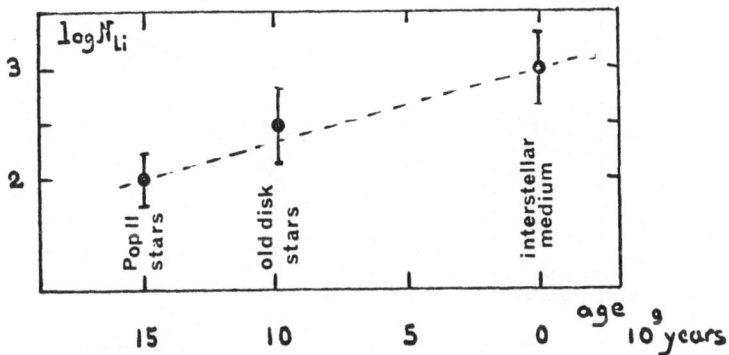

Figure 8. Observed lithium abundance as a function of the age of the Universe (from Spite and Spite [7].

Together with O. Boulade, G. Malinie and Y. Poilane, I am re-
investigating the evolution of ^{7}Li (along the lines similar to
those of Audouze and Tinsley [54]) to check if novae and/or red
giant stars could explain the significant increase in ^{7}Li noticed
in the Solar System and in the interstellar medium. We use a
production matrix displayed in figure 9 analogous to that used in
Alloin et al., [45] but adapted for the ^{7}Li production. We assume
that novae and/or red giants which are responsible for this
production evolve as low mass stars of 1 - 2M_Θ. We obtain the
results shown in figure 10 which are in good agreement with the
observations in both cases where the infall is or is not taken
into account. It is interesting to note that the Li overabundances in novae which are needed to account for the ^{7}Li observed
abundances are quite consistent with the Li upper limit determined by Friedjung, [55], which go from Li (nova)/Li$_\Theta$ = 6000 for
HR Del to 30000 for NQ Vulp. In our calculations if novae eject
on the average 10^{-3} M_Θ per event one needs Li(nova)/Li$_\Theta$ = 600

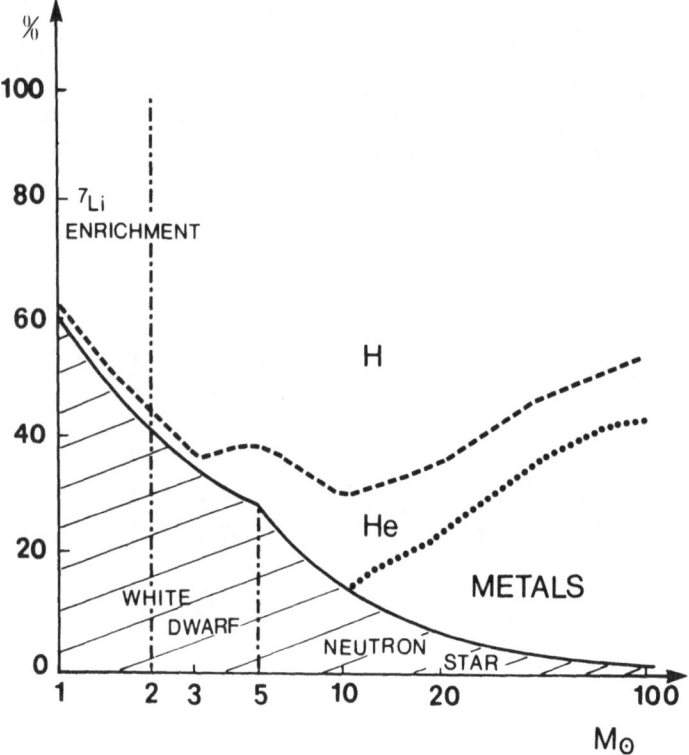

Figure 9.   On this figure derived from figure 7 the stellar mass
range responsible for the ^{7}Li enrichment is shown (according to
Audouze, Boulade, Malinie and Poilane).

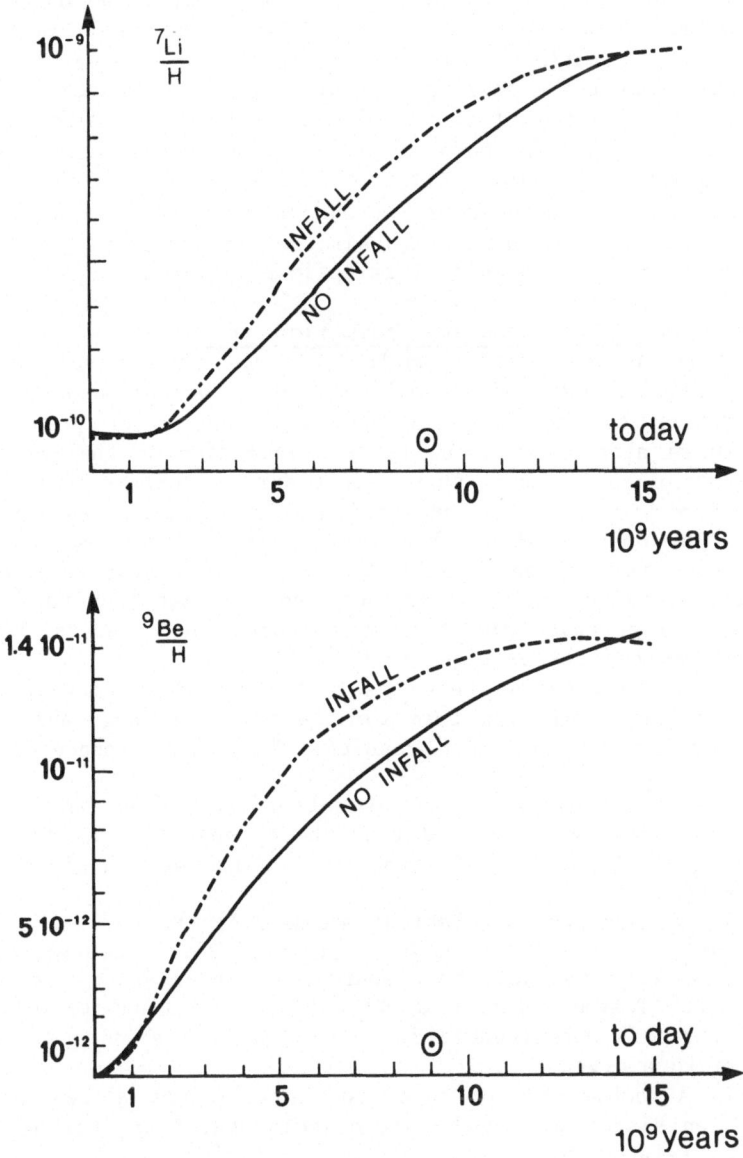

Figure 10.   Results of the calculations performed by Audouze, Boulade, Malinie and Poilane concerning the evolution of [7]Li and [9]Be. They use models with and without infall of external material. Both models reproduce satisfactorily the observations.

which should be 6000 if novae eject only $10^{-4} M_\odot$ on the average. Another possible test to decide if ^{7}Li comes from novae outburst is to look at the 0.478 MeV gamma ray line which can be emitted by the ^{7}Li desexcitation as proposed by Audouze and Reeves, [56].

About 1 % of red giants show significant enhancements in ^{7}Li (Li/H $\approx$ 2 10^{-6} instead of 10^{-9}). Therefore there are also valuable sources of ^{7}Li as shown by our current investigation and already suggested by Scalo, [57].

In brief, a low primordial ^{7}Li abundance which is consistent with standard Big Bang nucleosynthesis does not create any difficulty with the current models of galactic evolution.

4 - 2 Planetary nebulae and star formation rate.

In this section too I would like to mention some work currently carried out with G. Malinie and M. Dennefeld in which we are trying to use the abundance distribution of planetary nebulae in an attempt to deduce some information on the rate of star formation of low mass stars. In order to analyze this rate of star formation a large unbiased and homogeneous sample is required. The objects should have about the same age as that of the galactic disk and one should be able to determine this age. The assets of using the planetary nebulae are that they have low mass stars progenitors but they are conspicuous objects since they have luminosities as high as 10^3 - $10^4 L_\odot$.

I will only describe here the method we adopt to deduce a stellar formation rate from them for the low mass range and give some indication on our first results. This method proceeds in three steps :

(i) One uses the oxygen and/or sulfur abundances (the first ones are more safely determined than the second) to construct an histogram i.e. the number of planetary nebulae with a given metallicity (figure 11).

(ii) One uses the correlations age metallicity established by Twarog, [49] from the Fe/H abundance (figure 12). Of course there are at this stage two important sources of uncertainty : a) the fact that the Twarog relation applies to Fe/H and not to O/H or S/H ; b) there is a noticeable dispersion of Fe/H abundances corresponding to a given age.

(iii) A combination of these two correlations allows us to construct an histogram : number of objects-age. Since the age of a star is related to its mass by the relation $m(t) \approx 10^{10/3} t^{-1/3}$ where t is expressed in 10^9 years unit and m in solar unit, one can deduce an histogram : number of objects-mass (figure 13). This histogram may provide some limits on the rate of star formation and on the evolution of the initial mass function. It seems to favour (at the time when this review is being written) fairly steep stellar formation rates in the low mass range and for significant variations of the IMF with time. This preliminary result has to be discussed against the conclusion of Twarog [49] according to whom the rate of star formation in the

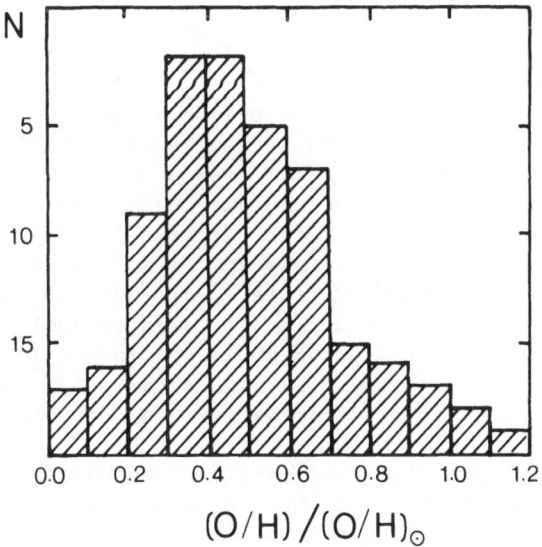

Figure 11. Histogram of a sample of planetary nebulae with respect to their oxygen abundance (Malinie et al. 1983).

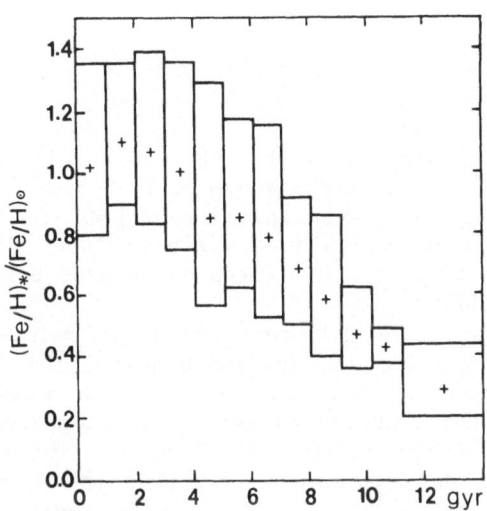

Figure 12. Distribution of the (Fe/H) abundance of the main sequence stars belonging to the galactic disk with respect to their
age (from Twarog, [49]).

past (i.e. at the formation of the galactic disk) should not have
a value more than   5 times the value of the present rate.

4 - 3 Effects of the infall on the dynamics and the chemistry of
         a spiral galaxy.

As has been observed in the previous sections, the possibility of infall of external gas may solve many problems : it is a
way to prevent too steep an increase of the metallicities with
time ; and it may explain the presence of deuterium in the galactic center, despite the fact that interstellar matter has been
very efficiently processed into stars in such a region. There is
other evidence in favour of infall such as the high velocity
clouds which, according to Oort [58], seem to fall on the galactic disk - although this statement is challenged for instance by
Verschuur [59]. In any case, Mayor and Vigroux [60] have examined
the effects of such infall processes in the dynamic and the chemical evolution of our galaxy.

They first show that when a galactic disk is submitted to
infall, the conservation of angular momentum implies that the
disk should contract : the radial velocity of matter inside the
galactic disk can be simply computed if we assume that the accretion velocity is parallel to the rotation of the axis, if the
velocity of the galaxy itself is low and if one neglects any
magnetic field effect. In these conditions the radial velocity of
the galactic disk matter is given by :

$$v_r = \frac{-\sigma_{acc}(r,t)}{\sigma(r,t)} \quad \frac{I(r)}{dI(r)/dt}$$

where σ_{acc} is the accretion (infall) rate, $\sigma(r,t)$ is the surface gas density, $I(r)$ and $dI(r)/dt$ are the angular momentum and
its first time derivative.

For infall rate equal to the rate of star formation
($\approx 5 \; 10^{-3} \; M_\odot \; kpc^{-2} \; yr^{-1}$) and $\sigma \approx 4 M_\odot \; pc^{-2} \; v_R \approx 10 \; km \; s^{-1}$.

One conclusion of Mayor and Vigroux [60] is that infall of
external gas implies inflow i.e. mixing between different zones
of the galactic disk. In other words one cannot build up one zone
models with infall.

They then analyzed different evolution models to follow up
the rotation curve, the gas surface density, the star formation
rate, the O/H abundance and the $^{12}C/^{13}C$ ratio along the galactic
disk. They consider that the galaxy has two components : one bul-
ge (radius r = 5 kpc, mass = 4 $10^{10} M_\odot$ and where $\sigma(r) \propto r^{-1/4}$
according to the de Vaucouleurs law) plus one disk (r = 7kpc at
t = 0 and mass = 7 10^{10} $M_\odot$). They divide their models into 25
concentric rings containing equal mass of material and having
radii from 0.4 to 20 kpc. They assume that the total mass and the
metal content are conserved. Table 6 provides the other characteristics of their four models. ν and K are the parameters entering in the Schmidt law : the rate of star formation σ_S (per

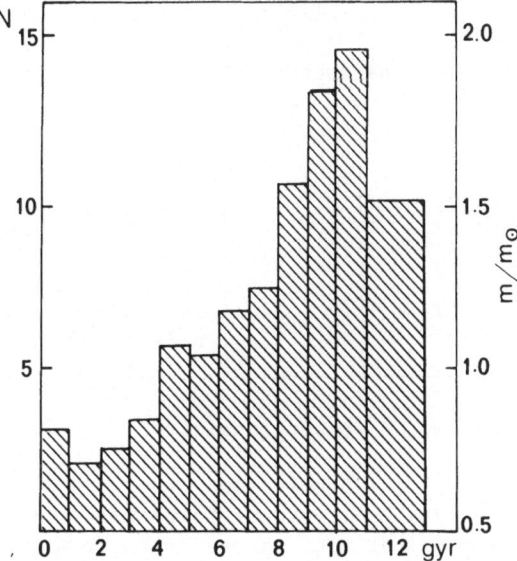

Figure 13. Histogram of the planetary nebulae (same sample as in figure 11) with respect to their age. This histogram shows that there are many more old planetary nebulae than young ones. This implies either that the rate of formation of low mass stars was bigger in the past and/or that the initial mass function has changed significantly during the galactic evolution (Malinie et al, 1983).

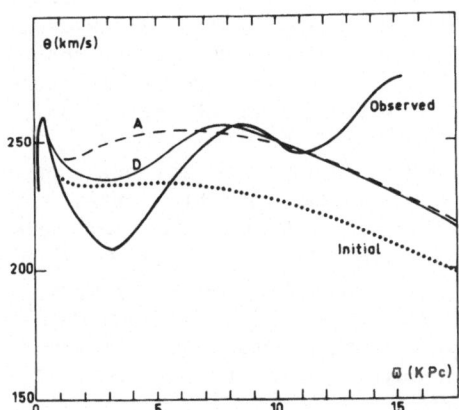

Figure 14. Rotation curve along the galactic disk : the heavy line corresponds to the observed rotation curve of our Galaxy. The dotted line corresponds to the rotation curve of our initial Galaxy and the dashed line and thin line to the rotation curves obtained in models A and D respectively at t = 12.4 10^9 years (Mayor and Vigroux, [60]).

TABLE 6

Values of the parameter entering in the model.

Models	ν	K	δ	v_r
A	0.03	2	0.01	yes
B	0.03	2	0.01	no
C	0.5	1	0.01	yes
D	0.03	2	r 5kpc : 0.0	
			r 5kpc : 0.01	

ν is the coefficient of proportionality of the Schmidt law for the SFR
K is the exponent of the Schmidt law
δ is the rate of infall in unit of total mass and 10^9 years
v_r indicated when radial motions are taken into account or not

========

surface unit) is given by $d\sigma_s/dt = v\sigma^K$; δ is the infall rate in units of 10^9 years per unit of total mass. One should note that in model B, they arbitrarily decide to shut off the radial inflow.

For the rotation curve as seen in figure 14, the best agreement is achieved by model D where infall concerns only the outer part of the galactic disk. Concerning the hydrogen (H I + H_2) surface density (figure 15) the agreement is reasonably good for all models but model D is the only one to reproduce the observed hole in the surface density of central regions with r $<$ 5kpc.

One cannot distinguish between models A,C, or D concerning the star formation rate (figure 16) and one can only note that model B leads to too low a star formation rate.
The inconsistency of model B (which does not take into account the radial inflow) is more obvious in the (O/H) dependence with radius while model D is still the one which most closely approaches the observed data (figure 17). Finally the $^{12}C/^{13}C$ ratios, coming from the H_2CO measurements, are displayed in figure 18 together with the results obtained from model A, which are in fairly good agreement.

In conclusion, the infall of external matter necessarily implies mixing between different zones of the galactic disk. Infall and inflow have direct influence on various processes or parameters such as the rate of star formation, the interstellar gas density and the abundance evolution. Model D in which infall operates only in outer zones with r $<$ 5kpc, might be the one which is the most consistent with observations.

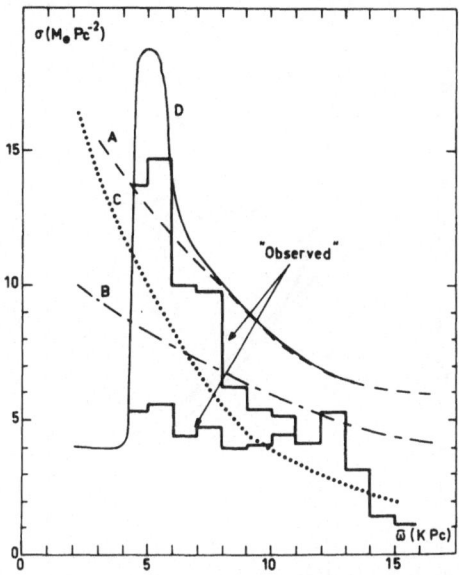

Figure 15. Gas surface density along the galactic disk. To determine the gas density one should make an assumption on the proportionality between CO and H_2. This ratio might be affected by a gradient in C/H and O/H. The upper heavy broken line corresponds to the gas density determination a constant CO/H_2 ratio. The lower heavy broken line is obtained by taking into account some gradient in O/H. The other lines correspond to the different models : A : dashed line ; B : dashed dotted line, C : dotted line; D : thin line (Mayor and Vigroux, [60]).

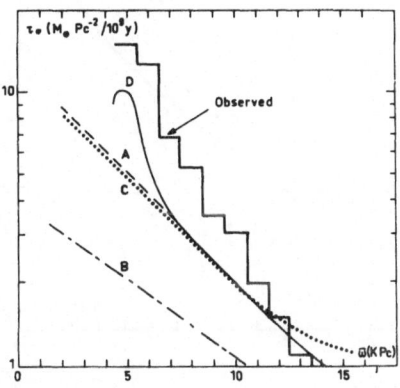

Figure 16. Star formation rate (SFR). The broken line corresponds to the observed SFR in our galaxy. The other lines correspond to the different models and the symbols are the same as in figure 15. (Mayor and Vigroux, [60])

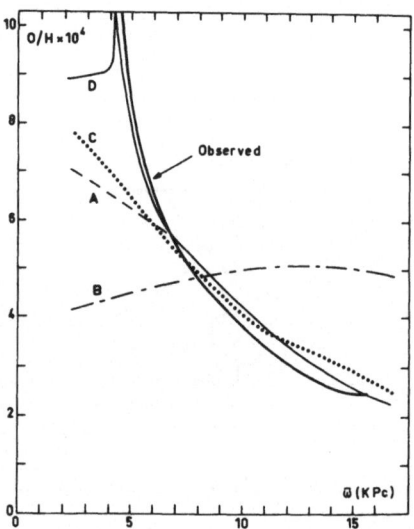

Figure 17. Oxygen abundance gradients. The heavy line corresponds to the abundances observed in H II regions. The other lines correspond to the different models and the symbols are the same as in figure 15. (Mayor and Vigroux, [60]).

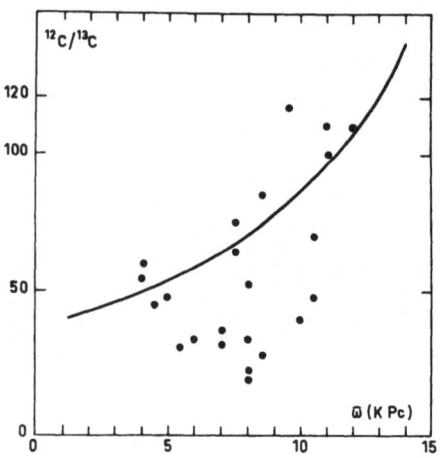

Figure 18. $^{12}C/^{13}C$ ratios. The dots correspond to the measured $^{12}C/^{13}C$ ratio from $H_2 CO$ observations in molecular clouds. The line corresponds to the prediction of Model A (Mayor and Vigroux, [60]).

4 - 4 Stochastic effects in the chemical evolution of galaxies.

In the preceding paragraphs we have made the reasonable assumption that the rate of star formation is directly related to the interstellar gas density. In an analysis performed with S. White [61] we investigate models in which there is no direct connections between this rate and this density. Moreover a given galactic zone is no longer assumed to have an homogeneous composition. Star formation does occur at random in any part of the considered region and the mixing of material ejected from stars with the interstellar matter is not complete. It should be realized in particular that the metallicity of a star no longer provides us with a determination of its age.

Nevertheless, to pin point the major effect of such models in the primary and secondary element distributions we have made the following simplifying assumptions :
i) all the newly synthetized elements are assumed to be returned to the interstellar medium instantaneously at star formation in finite quantities which are proportional to the mass of long lived stars and stellar remnants. ii) The yield of secondary elements is still assumed to be proportional to the abundance of primary elements in the protostellar gas while the yield of primaries is dependent of metallicity. iii) Each considered zone can exchange matter by infall and/or inflow.

The expanded version of our work presents some analytical approach and a Monte-Carlo simulation of this stochastic enrichment. Since the analytical formalism does not lead us to a fully consistent picture of the chemical evolution of the solar neighborhood. I restrict myself to a description of the "Monte-Carlo" approach. In this scheme, the stars and the interstellar gas are represented by a discrete set of mass units, each of which being either stellar or gaseous and where abundances are defined. An infall event is simulated by adding a gas unit with primordial abundances and a star formation event by converting a gas unit into a star unit. Mixing of infalling gas is modelled by taking n_{inf} gas units and reducing the metallicity of them by assuming it has been diluted with $1/n_{inf}$ of the infalling gas and at the same time replacing the metallicity by its average on $n_{inf} + 1$ units. Contamination is mimicked by choosing n_{con} gas units and adding y_1/n_{con} to the primary element abundances and $Y_2 Z_1^*/n_{con}$ to the secondary element abundances where Z_1^* is the primary abundance of the gas unit which has formed stars. Infall and star formation occur with a relative probability given by the rate function $R(M_S)$ such that $dMg/dM_S = R(M_S) - 1$.
In summary, this rate function $R(M_S)$ specifies the well-mixed infall model while its inhomogeneity is specified by the parameters n_{inf} and n_{con}, these describe the extent to which infalling gas and enriched stellar ejecta are mixed with the interstellar medium. Inhomogeneity in metal abundances is indeed observed in interstellar regions near the Sun. If n_{inf} and n_{con} are large they correspond to more classical homogeneous models.

In the models considered in the present study, one assumes that the infall should be a decreasing function with time ; and following Lynden Bell [62] one takes

$$R(M_S) = 2(1-M_o/M_\infty) (1- M_S/M_\infty)$$

where M_0 is the total mass when the galaxy starts to form stars, and M_∞ is the asymptotic total mass at later times. For Lynden Bell $M_0/M_\infty \approx 0.25$.

The Monte-Carlo simulations have been carried out by taking $M_\infty = 9000$ mass units and the results are displayed for $M_S = 8000$ mass units when $M_g/m_S = 0.1$. Figure 19 represents histograms of the primary metallicities for three models : when the inhomogeneity is large $n_{inf} = n_{con} = 3$, moderate $n_{inf} = n_{con} = 20$, and null $n_{inf} = n_{con} = \infty$. For all these models, $M_0/M = 0.25$ and $M_S/M = 0.88$, the yield has been chosen so that the mean value of log Z_1 corresponds to the mean value evaluated by Pagel and Patchett [43], for the nearby G dwarf distribution. Inhomogeneous models for which the mixing is intermediate between $n_{inf} = n_{con} = 3$ and $n_{inf} = n_{con} = 20$ seem to account for the histogram of the observed primary metallicities.

In figure 20 one compares the primary metallicity obtained in the frame of these two inhomogeneous models with the age metallicity relation proposed by Twarog [49]. We also find from this comparison that the best fit should be obtained by a model for which the extent of mixing is between those of the two considered models : the dispersion about the age metallicity relation is a factor 1.5 too small for the better mixed model and a factor 2 too large for the poorly mixed model.

The present study has also been motivated to explain the large dispersion observed in the secondary over primary abundance ratios. For instance, the nitrogen to oxygen abundance ratio shows no trend at all with the oxygen abundance. When one considers the abundance observations in H II regions, this is often taken as an argument in favour of a primary origin rather than a secondary one for this element. In fact as shown in figure 21 inhomogeneous models are quite consistent with the observed pattern even if nitrogen is assumed to be a purely secondary element. The conclusions are fairly similar for the N/Fe versus Fe/H trends coming from stellar abundances.

Finally, in figure 22 one can notice that the very strong dependence of Z_2/Z_1 with Z_1 predicted by an homogeneous model is much weaker when inhomogeneous models are used. Therefore the fact that there is not a strong correlation between [N/Fe] and [Fe/H] does not mean that N is not a secondary element.

Most of the hypotheses made to construct our "inhomogenous" models seem to be either plausible or (which has the same effect) critical : the instantaneous rejecting approximation is not very critical if the metallicity distribution evolves weakly over the stellar lifetimes involved. This would be questioned only if low mass stars play a significant nucleosynthetic role (this does not seem necessary for N which does not have to be produced as a pri-

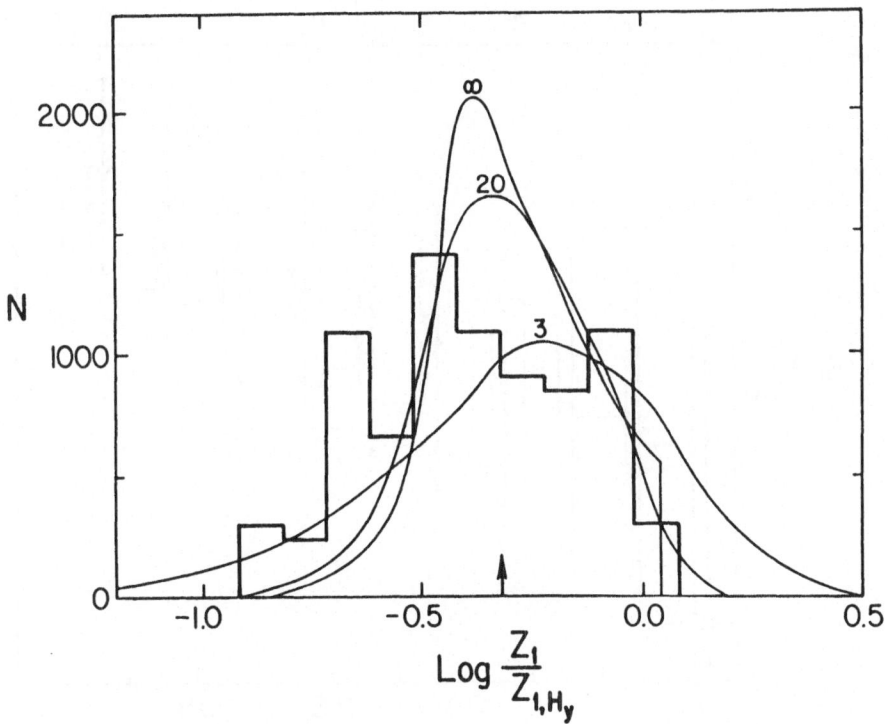

Figure 19. The local surface density of G-dwarf stars as a function of primary metallicity, taken from Pagel & Patchett [43], is compared with the predictions of three generalized Lynden Bell models. These models all have a $M_O/M_\infty = 0.05$ and $M_S/M_\infty = 0.88$ at the epoch plotted. The yield in the models has been chosen so that their mean value of Log Z, coincides with Log $Z_1/Z_{1,Hy} = -0.32$, the weighted mean of the values given by Pagel & Patchett. The models and the data have been renormalized to enclose equal areas, and the values for n_{inf} and n_{con} are equal in each model and have been used to label the curves. (White and Audouze, [61].

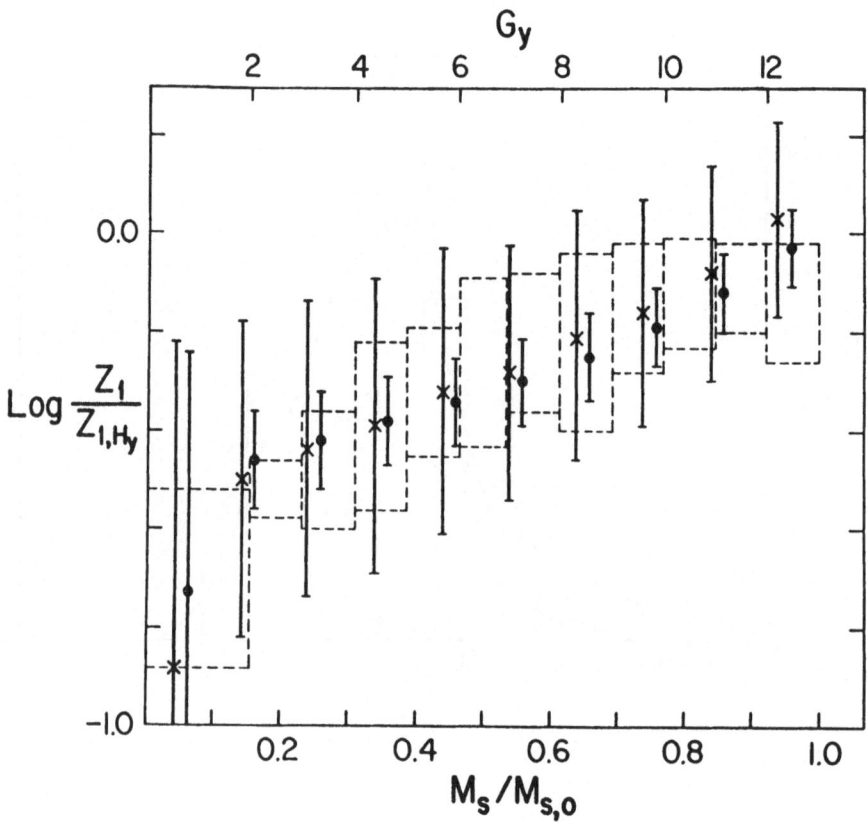

Figure 20. The points with error bars show the mean value and
the rms dispersion of logarithmic abundance as a function of time
of birth (as judged by a fraction of the final stars formed) for
the two inhomogeneous models of figure 19. The dashed boxes show
Twarog's preferred age-metallicity relation derived assuming a
variable helium abundance (Twarog, [49]) ; the height of each box
is twice the rms dispersion about the mean. The two horizontal
scales in this diagram have been plotted assuming a constant star
formation rate. (White and Audouze, [61]).

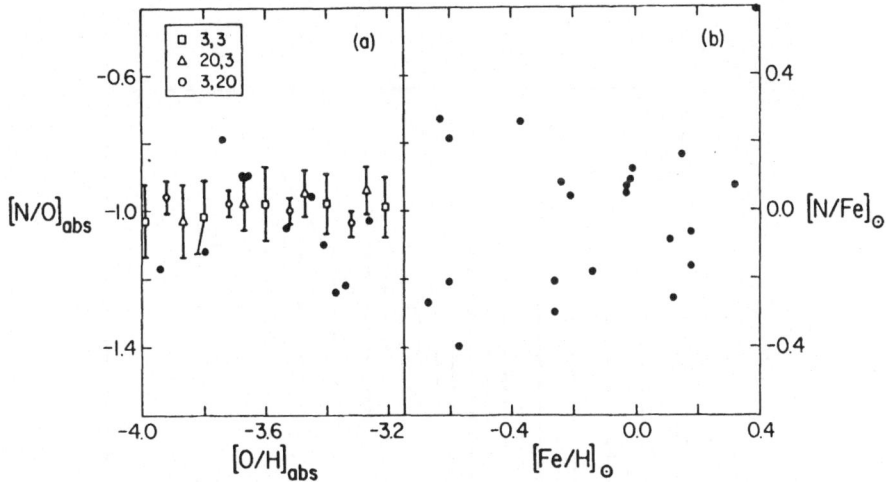

Figure 21. a) The relative abundance by number of nitrogen to oxygen is plotted as a function of the oxygen abundance relative to hydrogen for 11 galactic H II regions. The points with error bars show the predicted mean relation and rms dispersion for several inhomogenous models. The symbols are identified by the (n_{inf}, n_{con}) values of the models. (White and Audouze [61]).
      b) Values of the relative abundance of nitrogen to iron are plotted as a function of the absolute iron abundance for a number of nearby dwarf stars. (White and Audouze, [61]).

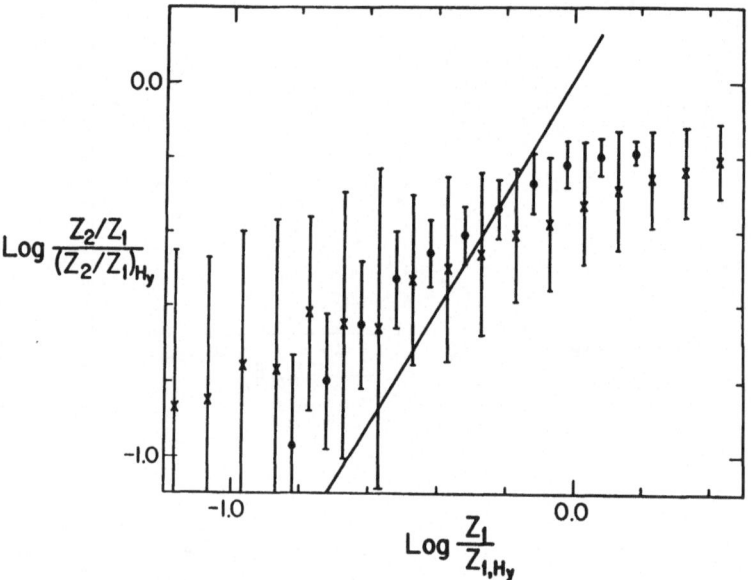

Figure 22. The relative abundance of primary to secondary nucleo-synthesis products is shown as a function of overall primary abundance for stars in the three generalized Lynden-Bell models of figure 19. The primary yield of the models is normalized as in figure 1, while the ratio of secondary to primary yields, Y_2/y_1, is the same and is arbitrary in the three models. (White and Audouze, [61]).

mary product of long lived stars) or if there is a strong initial burst of star formation.

Another assumption is that star formation is independent of metallicity. This assumption does not seem to suffer from any difficulty at present.

Besides the fact that N can be a bona fide secondary element, our models show that significant inhomogeneity takes place if the ejecta from stars are mixed with a mass of interstellar gas $\lesssim$ 20 times the total mass of stars formed or if the infalling material is mixed with $\lesssim$ 20 times their own mass of disk material. This hypothesis might be the most difficult to verify since it depends critically on the detailed nature of the motions in the interstellar medium. In any case, the results of our models should be valid if the metallicity of the interstellar medium is weakly correlated with the stellar metallicities on time scales comparable to the time scale for ejection of secondary elements which seems to be quite a plausible hypothesis.

In conclusion, these inhomogeneous models reproduce quite well the primary and secondary element abundance distribution and their dispersion such as the metallicity-age distribution observed by Twarog and are consistent with a secondary origin for N.

5 - Concluding remarks.

In this chapter, I have attempted to review the status of some current studies regarding the nucleosynthetic processes which are responsible for the formation of the observed nuclear species and the chemical evolution of galaxies. A few comments can be made at the end of this review.

- Concerning the nucleosynthetic processes, one does not expect very dramatic changes in the near future in the events which occur during the normal course of the stellar evolution or on the interaction between the cosmic rays and the interstellar medium. Significant progress is still expected in the studies of exploding objects (novae and supernovae). Although research in these quite complex fields seems to be carried out along profitable lines, a lot of work is still desperately needed to determine the proper equation of state of matter in any physical conditions and to link in satisfactory ways the formalisms which describe the thermodynamics, the hydrodynamics, the radiation transfers and the nuclear reaction effects on the exploding material. Concerning the primordial nucleosynthesis, the situation seems to be (at least at first sight) quite satisfactory. One has to keep in mind how-ever that too many hypotheses are made in the standard Big Bang models which seem to account so well for the D, ^{3}He, ^{4}He and ^{7}Li nucleosynthesis. In particular if another lepton is discovered, or if the Big Bang expansion has to be modified, such standard models may be found to be in difficulty.

- The construction of models of chemical evolution of galaxies is indeed an exciting exercise : it forces the model builder to link together various (and sometimes not very related) aspects of astrophysics such as abundance determinations, prediction of stellar formation rates and stellar mass distributions, etc... The results obtained so far, which are fairly satisfactory in many respects, should not induce us to forget that many of our models are only clever ways to parametrize our ignorance. More-over, the most significant progress regarding the understanding of the galactic evolution will only be made when it is possible to consider together the dynamical, the chemical and the morpho-logical evolution of galaxies.

In any case almost all of the questions touched upon in this review will benefit most by the launch of the Magellan observato-ry which should be the second generation mis'sion in the Coperni-cus class of projects. A better understanding of the primordial nucleosynthesis and the galactic evolution would be reached if one could estimate the interstellar abundances in the Magellanic clouds and in different regions of our Galaxy.

Acknowledgements.

I am indebted to Madeleine Steinberg, Marie-Claude Pantalac-ci and Françoise Warin who took care of the presentation of this manuscript, to Cecile Gry for suggestions on the draft of the paper and to Mary McClean who helped me to shape the style of the contribution.

REFERENCES

[1]   Audouze,J. and Vauclair,S. : 1980, an introduction to Nuclear Astrophysics, Reidel, Dordrecht
[2]   Biswas, S., Ramadurai,S. and Vahia,M.N. Eds. : 1981, Nucleo-synthesis, Tata Institute of Fundamental Research, Bombay
[3]   Wilkinson,D. Ed. : 1981, Nuclear Astrophysics, Pergamon Press
[4]   Barnes,C., Clayton,D.D. and Schramm,D.N. Eds. : 1982, Essays in Nuclear Astrophysics, Univ. Cambridge Press, Cam-bridge
[5]   Audouze,J. : 1982, in Astrophysical Cosmology Eds. H.A. Bruck, G.V. Coyne and M.S. Longair, Pontificiae Acade-micae Scientiarum Scripta Varia, City of the Vatican p. 395
[6]   Steigman,G. : 1982 in the Birth of the Universe Eds. J. Au-douze and J. Tran Thanh Van, Editions Frontières, Gif-sur-Yvette, p. 143
[7]   Spite,F. and Spite, M. : 1982, Astron. Astrophys. in press.
[8]   Kunth,D. : 1981, Ph.D. Thesis, Univ. Paris 7, see also in Cosmology and Particles, 1981, Eds. J. Audouze et al., p.245

[9] Rood,R.T., Wilson,T.L. and Steigman,G. : 1979, Astrophys. J. Letters 227, L 97

[10] Wagoner,R.V. : 1973, Astrophys.J. 179, 343

[11] Wagoner,R.V. : 1980, in Physical Cosmology, Eds. R. Balian et al., North Holland, p.398

[12] Yang,J., Schramm,D.N., Steigman,G. and Rood,R.T. : 1979, Astrophys. J., 227, 697

[13] Olive,K.A., Schramm,D.N., Steigman,G., Turner,M.S. and Yang. J. : 1981, Astrophys. J. 246, 557

[14] Sciama,D.W. : 1982, in Astrophysical Cosmology, Eds. H.A. Bruck, C.V. Coyne and M.S. Longair, Pontificiae Academicae Scientiarum Scripta Varia, p. 529

[15] Reeves,H., Fowler,W.A.W. and Hoyle,F. : 1970, Nature, 226, 727

[16] Meneguzzi,M., Audouze,J. and Reeves,H. : 1971, Astron. Astrophys. 15, 337

[17] Garcia Munoz,M., Mason,G.M. and Simpson,J.A. : 1975, Astrophys. J. Letters 201, L 141

[18] Reeves,H. and Meyer,J.P. : 1978, Astrophys. J. : 1978, Astrophys. J. 226, 613

[19] Burbidge,E.M., Burbidge,G.R., Fowler,W.A. and Hoyle,F. : 1957, Rev. Modern Phys. 29, 547

[20] Cameron,A.G.W. : 1957, Pub. Astron. Soc. Pacif. 69, 201

[21] Clayton,D.D. : 1968, Principles of Stellar Evolution and Nucleosynthesis, McGraw-Hill, New York

[22] Reeves,H. : 1964, Stellar Evolution and Nucleosynthesis, Gordon and Breach, New York

[23] Cameron,A.G.W. and Fowler,W.A. : 1971, Astrophys. J. 164, 11

[24] Langer,G.E., Kraft,R.P. and Anderson,K.S. : 1974, Astrophys. J. 189, 509

[25] Arnould,M. : 1982, Explosive Nucleosynthesis, Univ. Libre de Bruxelles

[26] Friedjung,M. Ed. : 1976, Novae and Related Stars, Reidel, Dordrecht

[27] Starrfield,S., Sparks,W.M. and Truran,J.W. : 1974, Astrophys.J. Suppl. 28, 247

[28] Audouze,J., Truran,J.W. and Zimmerman,B.A. : 1973, Astrophys. J. 184, 493

[29] Lazareff,B., Audouze,J., Starrfield,S. and Truran,J.W. : 1979, Astrophys. J. 228, 875

[30] Sneden,C. and Lambert,D.L. : 1975, Monthly Not. Roy. Astron. Soc. 170, 533

[31] Audouze,J. and Fricke,K.J. : 1973, Astrophys. J. 180, 239

[32] Wallace,R.K. and Woosley,S.E. : 1981, Astrophys. J. Suppl. 45, 389

[33] Vangioni-Flam,E., Audouze,J. and Chièze,J.P. : 1980, Astron. Astrophys. 82, 234

[34] Audouze,J., Chièze,J.P. and Vangioni-Flam,E. : 1981, Astron. Astrophys. 91, 49

[35] Arnould,M. and Norgaard,H. : 1981, Comments Astrophys. 9,145

[36] Arnould,M., Norgaard,H., Thielemann,F.K. and Hillebrandt,W.:
         1980, Astrophys. J. 237, 931
[37] Arnould,M. and Norgaard,H. : 1978, Astron. Astrophys. 64,
         195
[38] Weaver,T.A. and Woosley,S.E. : 1980, Ann. NY Acad. Sci. 336,
         335
[39] Audouze,J. and Truran,J.W. : 1975, Astrophys. J. 202, 204
[40] Woosley,S.E. and Howard,W.M. : 1978, Astrophys. J. Suppl.
         36, 285
[41] Audouze,J. and Tinsley,B.M. : 1976, Ann. Rev. Astron. Astro-
         phys. 14, 43
[42] Tinsley,B.M. : 1980, Fundamentals of Cosmic Physics. 5, 287
[43] Pagel,B.E.J. and Patchett,B.E. : 1975, Monthly Not. Roy.
         Astron. Soc. 172, 13
[44] Guélin,M. and Lequeux,J. : 1980, in Interstellar Molecules,
         Ed. B.H. Andrew, Reidel, Dordrecht
[45] Alloin,D., Collin-Souffrin,S., Joly,M. and Vigroux,L. : 1979
         Astron. Astrophys. 78, 200
[46] Wannier,P.G., : 1980, Ann. Rev. Astron. Astrophys. 18, 399
[47] Pagel,B.E.J. and Edmunds,M.G. : 1981, Ann. Astron. Astro-
         phys. 19, 77
[48] Lequeux,J. : 1979, Astron. Astrophys. 71, 1
[49] Twarog,B.A. : 1980, Astrophys.J. 242, 242
[50] Shaver,P.A., McGee, R.X., Newton,L.M., Danks,A.C. and Pot-
         tasch,S.R : 1982, preprint
[51] Vigroux,L., Audouze,J. and Lequeux,J. : 1976, Astron. Astro-
         phys. 52, 1
[52] Audouze,J., Lequeux,J., Rocca-Volmerange,B. and Vigroux,L.
         : 1976, in CNO isotopes in Astrophysics, Ed. J. Audouze
         Reidel, Dordrecht
[53] Audouze,J., Lequeux,J., Reeves,H. and Vigroux,L. : 1976, As-
         trophys. J. Letters 208, L51
[54] Audouze,J. and Tinsley,B.M. : 1974, Astrophys. J. 192, 487
[55] Friedjung,M. : 1979, Astron. Astrophys. 77, 357
[56] Audouze,J. and Reeves,H. : 1982, in Essays in Nuclear Astro-
         physics, Ed. C. A.Barnes et al., Univ. of Cambridge
         Press, to be published.
[57] Scalo,J.M. : 1976, Astrophys.J., 206, 795
[58] Oort,J.H. : 1974, in the Formation and the Dynamics of Ga-
         laxies, IAU, Symp. no. 58, Ed. J.R. Shakeshaft, Reidel
         Dordrecht
[59] Verschuur,G.L. : 1975, Ann. Rev. Astron. Astrophys. 13, 257
[60] Mayor,M. and Vigroux,L. : 1981, Astron. Astrophys. 98, 1
[61] White,S.D.M. and Audouze,J. : 1982, Monthly Not. Roy.
         Astron. Soc. in press
[62] Lynden Bell,D. : 1975, in Vistas in Astronomy 19, 299

DYNAMICS AND ENERGETICS OF THE INTERSTELLAR MEDIUM

Bernard Lazareff

Department of Astronomy
University of California, Berkeley
            and
Groupe d'Astrophysique / CERMO
Université de Grenoble

Some basic tools for the modeling of the dynamical properties of the diffuse interstellar medium are reviewed.   First, an
overview of fluid mechanics is given. Then, basic phenomena are
presented :   radiative losses, shocks, ionization fronts, and
electron thermal conduction. These tools and building blocks are
then applied to model idealized versions of localized dynamical
situations in the diffuse interstellar medium.   Some observational constraints on global models of the interstellar medium are
briefly discussed.

1.   CONTINUOUS FLUID DYNAMICS

The following is an overview of fluid mechanics directed to
applications to the interstellar medium.   Its scope will be
limited both by available space and the writer's competence.
Besides, a number of textbooks on that subject are available. We
shall start with a primer on the basic equations, that knowledge-
able readers may wish to   skip, then add comments and words of
caution on a few phenomena which are or may be relevant in
astrophysical situations.

We start by writing down the equations for plane parallel
flow in Lagrangian form:

$$\frac{D}{Dt}\rho = -\rho\frac{\partial}{\partial x}u \qquad\qquad (1.1a)$$

J. Audouze et al. (eds.), Diffuse Matter in Galaxies, 141–204.
Copyright © 1983 by D. Reidel Publishing Company.

$$\frac{D}{Dt}u = -\frac{1}{\rho}\frac{\partial}{\partial x}P \tag{1.1b}$$

$$\frac{D\varepsilon}{Dt} = -\frac{P}{\rho}\frac{\partial}{\partial x}u \tag{1.1c}$$

where:

$$\frac{D}{Dt} = \frac{\partial}{\partial t} + u\frac{\partial}{\partial x}$$

is the Lagrangian derivative, i.e. the rate of change seen by an observer riding along with the fluid, and ε is the internal energy per unit mass. Equations (1) become transparent if one imagines an element of fluid of unit mass, enclosed at constant t within a thin slice between x and x + dx, as follows its evolution to t + dt, noting that $\partial u/\partial x$ is just the rate of relative change V^{-1} dV/dt of any comoving volume of fluid.

Simple combinations of equations (1) yield the following equivalent set:

$$\frac{\partial \rho}{\partial t} = -\frac{\partial}{\partial x}(\rho u) \tag{1.2a}$$

$$\frac{\partial}{\partial t}(\rho u) = -\frac{\partial}{\partial x}(P + \rho u^2) \tag{1.2b}$$

$$\frac{\partial}{\partial t}(\rho\varepsilon + 1/2\ \rho u^2) = -\frac{\partial}{\partial x}[\rho u(1/2\ u^2 + \varepsilon + P/\rho)] \ . \tag{1.2c}$$

These equations are in Eulerian form. The quantities whose time derivatives appear on the left hand side are the <u>volume densities</u> of mass, momentum, and total (internal + kinetic) energies; the r.h.s. are the space derivatives of the <u>fluxes</u> of mass, momentum and energy. To visualize these equations, imagine a thin slab fixed in the "lab" frame, with mass, momentum, and energy flowing in and out through the walls. In particular, in the r.h.s. of (2c), the first and second terms are the kinetic and internal energy carried along (advected) by the fluid, while the third is the work done by pressure : P.u . The sum:

$$\frac{1}{2}u^2 + \varepsilon + P/\rho = h + \frac{1}{2}u^2$$

is called the stagnation enthalpy; in a stationary flow it is a conserved quantity equal to the enthalpy at a point (if any) where the fluid comes to rest.

The choice of one form of another of the equations depends

on circumstances. The Lagrangian form has the advantage that, since it "rides with the fluid" the somewhat trivial advection terms of the form u.dA/dx do not appear. On the other hand, the Eulerian form emphasizes conservation laws, which may add to one'e peace of mind in numerical work. Also, they can be integrated over a finite volume to yield equations of the form:

$$\int_a^b \partial G/\partial t = F(a) - F(b) \tag{1.3}$$

where the time derivative can be pulled out of the integral if the boundaries are fixed. In that form (which we could háve derived directly from conservation laws) the equations are valid irrespective of the details of the flow within the "budget volume" a < x < b; the flow may even exhibit spatial discontinuities. Equations (2) cast in the form (3) are especially handy when studying a stationary flow.

For completeness, we write down the general form of the equations in the Lagrangian version:

$$\frac{D}{Dt}\rho = -\rho \; \underset{\sim}{\nabla} \cdot \underset{\sim}{u} \tag{1.4a}$$

$$\frac{D}{Dt}u = -\frac{1}{\rho} \underset{\sim}{\nabla} \cdot P \tag{1.4b}$$

$$\frac{D}{Dt}\varepsilon = -\frac{P}{\rho} \underset{\sim}{\nabla} \cdot \underset{\sim}{u} \tag{1.4c}$$

and in the Eulerian version:

$$\frac{\partial \rho}{\partial t} = -\underset{\sim}{\nabla} \cdot (\rho \underset{\sim}{u}) \tag{1.5a}$$

$$\frac{\partial}{\partial t}(\rho \underset{\sim}{u}) = -\underset{\sim}{\nabla} \cdot (P + \rho \; \underset{\sim}{u} \otimes \underset{\sim}{u}) \tag{1.5b}$$

$$\frac{\partial}{\partial t}(\rho \varepsilon + \frac{1}{2}\rho u^2) = -\underset{\sim}{\nabla} \cdot [\rho \underset{\sim}{u}(\varepsilon + \frac{1}{2}\rho u^2) + \underset{\sim}{u} \cdot P] . \tag{1.5c}$$

Before proceeding, we must make explicit the simplifying assumptions underlying equations (1,2,4,5), which we left unmentioned at the outset in order not to clutter the basic argument.

-No creation or destruction of particles; breaks down when treating a multicomponent gas (with, for instance, ionization) or a multiphase flow (with, for instance, evaporation-/condensation mass exchange between phases). Affects all four equations (a), and possibly the other equations as well if matter is introduced into the flow with nonzero momentum or energy in the fluid (eqs. 1 and 4) or lab (eqs. 2 and 5) rest frame.

–No external forces (e.g. gravity, radiation pressure, drag against another cospatial fluid). Affects all equations (b), (2c) and (5c), but not (1c) and (4c).

–energy changes only from pdV at boundaries of fluid element, leaving out energy transport (conduction) or injection (radiative heating and cooling). Affects all equations (c).

–Scalar (isotropic) pressure.

We do not write down a fully general version of the equations with catchall extra terms; we think that one can deal with individual cases once one has understood the physical meaing of the basic equations. Specific examples will come up further in the text.

The derivation of the basic fluid dynamic equations can be found in many textbooks, e.g. [1], [2], [3], [4].

We shall now take an alternative route to equations (5), which brings up naturally the extant terms and some more. The derivation of the gas dynamic equations from the Boltzmann equation is treated in many textbooks on the kinetic theory of gases, e.g. [5] or [6]. We give below a minimal version of that derivation. We start with the collisionless Boltzmann equation, with no external forces:

$$\frac{\partial f}{\partial t} + v_x \frac{\partial f}{\partial x} + v_y \frac{\partial f}{\partial y} + v_z \frac{\partial f}{\partial z} = 0 \qquad (1.6)$$

where $f(r,v,t)dr\,dv$ is the number of particles in a small volume of 6-dimensional phase space. To see where (6) comes from, draw a small element of 3-space $\delta x \delta y \delta z$ and focus attention to particles having velocity δv about v. In the absence of collisions or external forces, the change of the population is given by the particle fluxes through the walls. For the yz walls at x and x + δx, the net flux is:

$$[v_x f(x,y,z,v,t) - v_x f(x + \delta x,y,z,v,t)]\delta v \ .$$

Note that v is an independent variable, not a position dependent mean velocity. Adding up the two other similar terms yields equation (6). We now multiply (6) in turn by 1, v, $(1/2)v^2$ and integrate over velocity space. Let:

$$n = \int f \, d\underset{\sim}{v} \qquad\qquad\qquad \text{particle density} \qquad (1.7)$$

$$\underset{\sim}{u} \; n^{-1} \int \underset{\sim}{v} \; f \, d\underset{\sim}{v} = \langle \underset{\sim}{v} \rangle \qquad\qquad \text{mean velocity} \qquad (1.8)$$

$$\underset{\sim}{w} = \underset{\sim}{v} - \underset{\sim}{u} \; , \; \langle \underset{\sim}{w} \rangle = 0 \qquad\qquad \text{"random" velocity} \qquad (1.9)$$

The zero order moment yields the number conservation:

$$\frac{\partial n}{\partial t} + \underset{\sim}{\nabla} \cdot (n\underset{\sim}{u}) = 0 \; . \qquad\qquad\qquad (1.10)$$

The first order moment in the x component is:

$$\frac{\partial}{\partial t} \int v_x \, f d\underset{\sim}{v} + \frac{\partial}{\partial x} \int v_x^2 \, f d\underset{\sim}{v} + \frac{\partial}{\partial y} \int v_x v_y \, f d\underset{\sim}{v} + \frac{\partial}{\partial z} \int v_x v_z \, f d\underset{\sim}{v} = 0$$

$$(1.11)$$

$$\frac{\partial}{\partial t}(n u_x) + \frac{\partial}{\partial x} n(u_x^2 + \langle w_x^2 \rangle) + \frac{\partial}{\partial y} n(u_x u_y + \langle w_x w_y \rangle)$$

$$+ \frac{\partial}{\partial z} n(u_x u_z + \langle w_x w_z \rangle) = 0 \qquad (1.12)$$

Gathering the three components into compact notation and multiplying by the mass m of individual particles, we obtain:

$$\frac{\partial}{\partial t}(\rho\underset{\sim}{u}) + \underset{\sim}{\nabla} \cdot (\rho\underset{\sim}{u} \times \underset{\sim}{u} + \underset{\approx}{P}) = \underset{\sim}{0} \qquad\qquad (1.13)$$

where:

$$P_{ij} = \rho \langle w_i w_j \rangle.$$

We see that the two terms in (2b) and (5b) both arise from the momentum carried by individual particles, and are split only because we have singled out the mean (hydrodynamical) velocity. We also see that pressure is in general a symmetric, but non-isotropic tensor. Without going through details, we give the result of integrating after multiplying by $(1/2)v^2$:

$$\frac{\partial}{\partial t}\left(\rho\varepsilon + \rho\frac{u^2}{2}\right) + \underset{\sim}{\nabla} \cdot \left[\rho\underset{\sim}{u}\left(\varepsilon + \frac{u^2}{2}\right) + \underset{\sim}{u} \cdot \underset{\approx}{P} + \underset{\sim}{q}\right] \qquad (1\cdot14)$$

where:

$$q_i = \rho \langle w_i w^2 \rangle$$

is an energy flux term which does not vanish in the fluid rest frame, and is to be identified with the heat flux. What if we now start over from (6), but with the collision term on the r.h.s. of (6)? We might show in detail that their contributions to the moment equations of order zero (10), one (13), and two (14), vanish; but this should be clear as long as particle number, momentum, and energy are conserved by collisions. Actually, we also require that the interparticle forces responsible for collisions be of short range (perfect gas limit), so that they do not contribute significantly to the stress and energy density of the fluid.

Collisions are in fact vital for establishing the equilibrium distribution of velocities (see references [5] and [6]). For a strictly Maxwellian distribution of velocities, the off-diagonal components of P, as well as q, vanish. For "small" deviations from thermodynamic equilibrium, one expects:

$$P_{ij}/P_{ii} = O(\lambda/L) \qquad i \neq j$$

$$q_i/\rho\epsilon^{3/2} = O(\lambda/L)$$

where λ, the mean free path between collisions, is the mean distance from which "rogue" particles have come to perturb the local equilibrium distribution, and L is a typical scale over which the thermodynamical quantities ($\rho, \epsilon \ldots$) change. The ratio:

$$Kn = \lambda/L$$

is called the Knudsen number of the flow, and characterises the transition between the collisionless regime (Kn $\gg$ 1, stellar systems) and the hydrodynamical regime (Kn $\ll$ 1, everyday gas dynamics).

From now on, we assume, unless otherwise stated, that we deal with a monoatomic gas in the hydrodynamical regime. We then have the closure relationship:

$$P = \frac{1}{3} \Sigma P_{ii} = \frac{2}{3} \rho\epsilon . \qquad (1.15)$$

Kinetic theory (resp. experiment) allows us to compute (resp. measure) the "small" transport terms:

$$\underset{\approx}{P} - p\underset{\approx}{I} = \mu \mathcal{L}(\partial u/\partial x) \qquad (1.16)$$

$$q = -k\nabla T \qquad\qquad\qquad (1.17)$$

where $p\underset{\sim}{I}$ is the scalar mean pressure; $\mathcal{L}(\partial u/\partial x)$ a loose notation for some linear combination of the spatial derivatives of u; μ the viscosity; k the heat conductivity; and $T = 2\varepsilon/3k_B$ the usual temperature. The system (10+13+14) together with equations (15) to (17) is now closed and is called the Navier-Stokes equations. Two dimensionless numbers are useful in discussing the nature of the solution of the Navier-Stokes equations.

The Mach number:

$$M = u/(5p/3\rho)^{1/2} = u/(10\ \varepsilon/9)^{1/2} \qquad\qquad (1.18)$$

characterizes the ratio of energies in ordered ($u^2/2$) and random (ε) motions. The numerical coefficient is such that the denominator is the speed of propagation of small adiabatic perturbations.

The Reynolds number:

$$Re = \frac{uL\rho}{\mu}$$

where as before, L is a characteristic scale of the flow. Rearranging:

$$Re = \frac{u^2\rho}{\mu(u/L)} \qquad ; \qquad u/L \approx \frac{\partial u}{\partial x}$$

we see that Re measures the ratio between momentum transport due to bulk motion and to viscous stresses. Since a naive gas-kinetic argument leads to:

$$\mu/\rho \approx (1/3)\varepsilon^{1/2}\lambda$$

we see that the previously defined numbers are not independent:

$$Re \approx M/Kn$$

The importance of M and Re stems from the fact that, if one scales chararcteristic masses, dimensions, etc., the scaled flow will be a solution of the scaled problem provided M and Re are kept invariant. In other words, for a given geometry, these two numbers determine the character of the flow. Therefore, their value should be checked when a) analyzing a given hydrodynamical situation; b) carrying results over from the vast body of "everyday life" hydrodynamic data to astrophysical situations; c) assessing the value of numerical simulations. In that respect, an overview of the problem of flow around a sphere or a cylinder is instructive. Figure (1.1) sketches various regions of the Re–M plane. The description of the flow in regions a through d can be found in most hydrodynamics textbooks, e.g. [7],[8], or [9]. A quantity often needed is the drag force, expressed as:

$$D = C_D \ 1/2\rho S u^2$$

where S is the area perpendicular to the flow, and C_D is called the drag coefficient. In region a, we have viscous (Stokes) flow, and $C_D = GM/Re$, so that $D \propto u$. As Re increases into region b, inertial forces start to dominate over viscous stresses. Around Re = 40, the flow becomes unsteady, with periodic

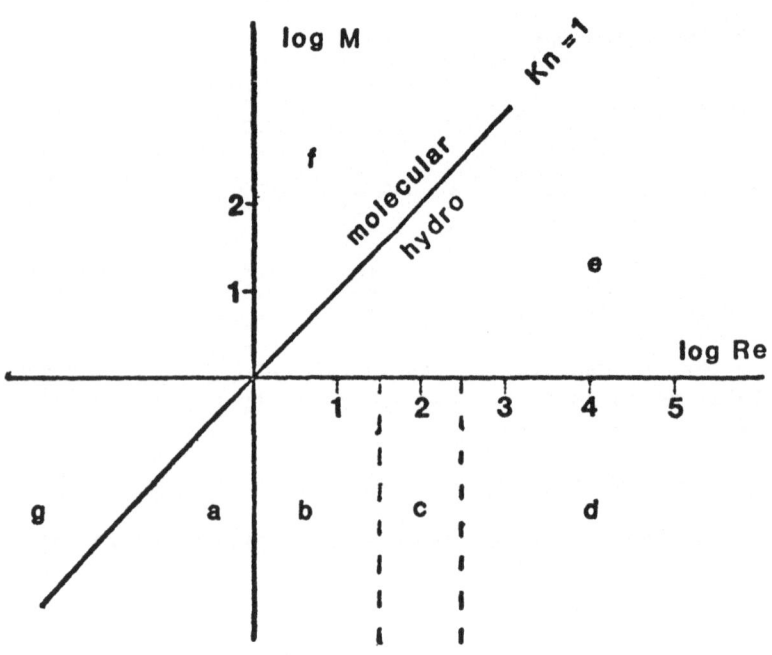

Fig 1.1 . The various flow regimes in the Reynolds versus Mach plane.

oscillations (region c); around Re = 200, the oscillations become chaotic: turbulence sets in (region d). The flow pattern near the surface of the body (sphere or cylinder) undergoes one last change around Re = 10^5. Despite this variety of flow regimes, the drag coefficient is essentially constant ($C_D \approx 0.4$ for a sphere) throughout all the region Re > 10, M < 1.   Region e is hydrodynamic supersonic flow; a shock develops ahead of the sphere, and $C_D \approx 1$ through that region. Finally, we have subsonic (resp. supersonic) molecular flow in region g (resp. region f). In region f, $C_D \approx 2$, while in region g the drag force is best expressed as:

$$D = kS\tau U$$

where τ is the 1-dimensional thermal velocity dispersion, and $k \approx 2-3$ depending on whether one assumes specular or diffuse reflection at the surface. A detailed treatment of regions f and g can be found in [10], and results for the transition between the molecular and the hydrodynamic regime are given in [11].

What is the relevant hydrodynamical regime in common astrophysical situations? Consider first a cloud (spherical, of course) of radius r = 3 pc moving through a neutral intercloud medium (n_{ic} = 0.1 cm^{-3}, T_{ic} = 10^4k) at velocity u = 7 km s^{-}. The Mach number is near sonic: M = 0.7. The collision mean free path $\lambda \approx 10^{16}$ cm and the Reynolds number: Re $\approx$ 1000. We are therefore in the hydrodynamic, subsonic, turbulent regime. Three important comments are called for at this point. First, none of the hydrocodes used for interstellar gas dynamics has a resolution as small as the Knudsen number (Kn ~ 10^{-3}) of this situation, and the complex transport properties at the boundary layer between cloud and intercloud cannot be modeled properly. Unless special precautions are taken, the finite resolution of a code introduces a numerical viscosity much larger than the actual one, and this will affect phenomena such as ablation through the Kelvin-Helmholtz instability. A second comment is that one occasionally finds in the literature the argument that the motions of a population of clouds in an external medium are naturally regulated at or near M = 1, under the implicit assumption that drag forces should rise dramatically in the supersonic regime. We have seen that C_D only increases by roughly a factor of 2 between the subsonic and supersonic regimes, hardly a large jump by astrophysical standards. The third comment is that the actual Reynolds number is often essentially unknown. In an ionized intercloud medium, the Coulomb mean free path can be comparable to the size of the object considered (cloud in supernova remnant, galaxy in intracluster gas), thus making viscosity dominant. The magnetic field, however, can modify this in two ways: a) a very small (even by astrophysical standards) magnetic field can reduce the effective mean free path of ions

much below the Coulomb value, thus reducing the viscosity; b) for
typical values of interstellar magnetic fields, the magnetic
stresses are comparable to the gas pressure, and are non-
isotropic; they can thus contribute to increase the effective
viscosity. Another illustrative example is the motion of a dust
grain relative to the gas, as might occur under the effect of
radiation pressure or behind a shock. The size of the grain is
much smaller than the gas mean free path, and the appropriate
regime is molecular flow, subsonic or supersonic as the case may
be. Formulae for the drag force are given in [10], while the
case of charged grains in an ionized gas is discussed in [12].

Before we leave the subject of fluid dynamics, we wish to
call the reader's attention upon yet another form of the
hydrodynamical equations : the characteristic form. A clear and
detailed exposition of equations in characteristic form is given
in § 5-8 of [2], with applications in § 9-12. Therefore, we
shall make only a few specific comments.
The characteristic equations can be used for any fluid
having a barotropic ($p = f(\rho)$) equation of state ; adiabaticity
is not required. In particular, characteristic equations can be
for an isothermal gas ($p = c^2\rho$); one cannot simply set $\gamma = 1$ in
the general equations, rather, one should start back from the
general expression :

$$J_{\pm} = u \pm \int dp/_{\rho} = u \pm c \, \mathrm{Log}(\rho) \qquad (1.19)$$

We also wish to present an alternative derivation of the
characteristic equations (for which we are unfortunately unable
to give proper credit, for lack of memory). Given a barotropic
equation of state, u and ρ completely specify the state of the
fluid, and only the first two (continuity and momentum) equations
are relevant. Now ask the following question : do the flow
equations allow the existence of a solution such that u and ρ
take prescribed values at nearby points (x,t) and $(x+\delta x, t+\delta t)$.
For this to be the case, the set of four partial derivatives of
(ρ,u) with respect to (x,t) must obey four equations : the two
flow equations of continuity and momentum, and the two constraints resulting from prescribed changes in ρ and u.

$$\partial\rho/\partial t + u \, \partial\rho/\partial x + \qquad\qquad \rho \, \partial u/\partial x = 0$$

$$c^2 \, \partial\rho/\partial x + \rho \, \partial u/\partial t + \rho u \, \partial u/\partial x = 0$$

$$\delta t \, \partial\rho/\partial t + \delta x \, \partial\rho/\partial x \qquad\qquad\qquad = \delta\rho \qquad (1.20)$$

$$\delta t \, \partial u/\partial t + \delta x \, \partial u/\partial x = \delta u$$

This linear system for the unknowns $\partial(\rho,u)/\partial(x,t)$ has, in general, a unique regular solution, except when the determinant vanishes :

$$\Delta = [(\delta x/\delta t)-(u+c)][(\delta x/\delta t)-(u-c)] \qquad (1.21)$$

If $\Delta = 0$, a solution can be found only if the right hand side of (1.20) satisfies a compatibility condition :

$$c \, \delta\rho \pm \rho \, \delta u = 0 \quad (\text{for } \delta x = (u\pm c) \, \delta t) \qquad (1.22)$$

Thus there are two directions in the (x,t) plane along which changes in ρ and u are tied together; this is expressed by the conservation of the Riemann invariants along their respective characteristics.

Characteristic equations are not used often for numerical calculations, but they are a powerful tool for understanding and qualitatively analyzing a flow, using notions such as : domain of influence, domain of dependance, and simple waves.

2. ENERGY SINKS

We shall discuss in this section the local energy loss mechanisms. Energy transfer by conduction will be discussed in section 5, and specific heating mechanisms in section 8. All the processes discussed in the present section involve collisions between particles followed by photon emission. We shall assume:
   a1) That once a photon is emitted, its energy is lost for gas. A sufficient, but not necessary condition is that the gas be optically thin to all emitted photons.
   a2) That electrons and ions share a common Maxwell-Boltzmann distribution of microscopic kinetic energy at some temperature T.
By convention we term "internal energy" the microscopic kinetic energy of gas particles; excitation and ionization energies are considered "external" to the system in that kind of book-keeping. The reason for that choice is that, once we become interested in the dynamics, we can use a simple relation between pressure p and internal energy per unit volume E: $p = (\gamma-1)E$. Situations where H_2 is abundant (see T. de Jong's contribution) may require special attention if the timescale for thermalization of H_2 rotational energy is not short compared to the timescale for temperature changes; the use of a simple algebraic relation between p and E is then not allowed. It might then be convenient to treat the H_2 rotational energy as external.

2.1 Atomic Collisional Losses: General Expression

Changes in the internal energy (as defined above) result from inelastic collisions between particles. For equal kinetic energies, an electron is more efficient than an ion in exciting an atomic transition with an energy jump ΔE of order kT, because its collision timescale is better matched to the atomic time-scale. Therefore, collisions involving electrons will dominate the cooling rate, unless their abundance is low; then one must consider the next lightest (and most abundant) species: hydrogen. The other partner of a collision can be any ion. We restrict for a moment the discussion to just one ionic stage "s" of one element "x," and to one bound excited level "ℓ" having an energy $h\nu$ above the ground state; and write down the corresponding contribution to collisional losses:

$$-(dE/dt)_{xs1} = n[n_e c_e(T) + n_H c_H(T)]h\nu$$

$$-n^*[n_e d_e(T) + n_H d_H(T)]h\nu$$

$$+nn_e[\alpha(T)\langle\varepsilon_r\rangle + i(T)\chi] \qquad (2.1)$$

where:

c_e, c_H rate coefficients $\langle\sigma v\rangle$ for collisional excitation
d_e, d_H rate coefficients for collisional de-excitation
n^* number density of xs ions in the excited state
α recombination rate coefficient for $s \to s-1$
         (radiative and dielectronic)
$\langle\varepsilon\rangle$ mean energy of recombining electrons ($\approx 3/2\ kT$)
i rate coefficient for collisional ionization
         $s \to s+1$
χ ionization energy for $s \to s+1$

Note that we have left out collisional ionization by H; is it safe, and when? We now proceed tentatively to compute the total radiative losses. Anticipating on the next subsection, we neglect collisional de-excitation. We restore indices x,s,ℓ:

$$\lambda_{e,xs} = \sum_{\ell} c_{e,xs\ell} + (\alpha\langle\varepsilon\rangle e i\chi)_{xs}$$

$$\lambda_{H,xs} = \sum_{\ell} c_{H,xs\ell}$$

$$-(dE/dt) = \sum_{xs\ell} (n_H n_{xs} c_{H,\ell} + n_e n_{xs} c_{e,\ell})$$

$$= n_H^2 \left[\sum_x \frac{n_x}{n_H}(\sum_s \frac{n_{xs}}{n_x} \lambda_{e,xs}) + \frac{n_e}{n_H}\sum_x \frac{n_x}{n_H}(\sum_s \frac{n_{xs}}{n_x} \lambda_{e,xs})\right]$$

$$= n_H^2 \left[\Lambda_H(T,A,I) + x\Lambda_e(T,A,I)\right] \qquad (2.2)$$

where:

$x = n_e/n_H$ ionized fraction
A   the set of element abundances in gas phase n_x/n_H
I   the set of ionization structures of all elements
          n_{xs}/n_x

We finally add free-free (bremsstrahlung) losses:

$$-(dE/dt)_B = n_e n_H \, b(T) = 2.4 \, 10^{-27} \, n_e n_H \, T^{1/2}$$

This numerical value is for cgs units, assuming a fully ionized gas (which will be the case anyway when bremsstrahlung is significant). The Gaunt factor has been taken from [3] at a temperature of 3.10^7 k, at which bremsstrahlung starts to dominate; it is a weak function of temperature.

Before we can actually compute the radiative losses, we must obtain, derive, or assume:
a3) relevant coefficients: mostly, but not always, known with sufficient precision.   a4) gas phase abundance: depends on both total interstellar abundances and depletion onto grains.   a5) ionization structure of each element: depends both on atomic physics and physical conditions (possibly past history as well) of the gas. Reasonable estimates can be made concerning a3 and a4; three distinct regimes leading to specific ways of obtaining the ionization structure, will be described in subsections 2.3, 2.4 and 2.5.

2.2 Collisional De-excitation

When are superelastic collisions important?   Write down the equation of statistical equilibrium for transitions between ground state and level $\underline{\ell}$:

$$n n_a c_a = n^* n_a d_a + n^* A \qquad (2.3)$$

where a = e or H, whichever dominates, and A is the Einstein coefficient for spontaneous emission so that the ratio of downwards to upwards collisional transitions is:

$$D = (n^* d_a)/(n c_a) = n_a/(n_a + A/d_a) \qquad (2.4)$$

The radio D is $\ll 1$ for $n_a \ll n_{a,crit}$ where:

$$n_{a,crit} = A/d_a \qquad (2.5)$$

and $D \simeq 1$ for $n_a \gg n_{a,crit}$.

The   net   rate   of   collisional   excitation:

$$nn_a c_a - n^* n_a d_a = n^* A$$

$$\simeq \begin{cases} nn_a c_a & (n_a \ll n_{a,crit}) \\ n(w^*/w)\exp(-h\nu/kT)A & (n_a \gg n_{a,crit}) \end{cases}$$

In the first limiting case—low density—all collisional excitations are followed by radiative decay, and the net rate is $\propto \rho^2$.
In the high density case, collisions establish a thermal population ratio, and the net rate is $\propto \rho$. In typical diffuse
interstellar medium situations, collisional de-excitation is
important for transitions with low values of A, "forbidden"
transitions; the collisional rates for these transitions, however
are comparable to those of "permitted" lines, hence they have
much lower n_{crit}. Note that whenever collisional de-excitation
is negligible, the population ratio n^*/n is small, because

$$n^* n = D. \; c_a/d_a = D. \; w^*/w \; \exp(h\nu/kT) \tag{2.6}$$

and $h\nu/kT$ is typically a few when a transition is a significant
contributor to energy losses. Is it safe to assume statistical
equilibrium to write (2.3)? Why and when? Is it safe to neglect
induced up and down radiative transition? The 21-cm line of H is
a "forbidden" transition for which induced radiative transtions
are important (see J. Lequeux's contribution). Does that invali-
date our argument? Why?

As an example, we compute the critical density for the
$^2P_{3/2} - ^2P_{1/2}$ transition of C^+, a major coolant in diffuse
neutral clouds, using collision strengths from [3]

$A = 2.4 \; 10^{-6} \; s^{-1}$

T (k)	de cm^{+3} s^{-1}	$n_{e,cr}$ cm^{-3}	d_H cm^{+3} s^{-1}	$n_{H,cr}$ cm^{-3}
10^4	$2.9 \; 10^{-8}$	83		
10^2	$2.9 \; 10^{-7}$	8.3	$8.0 \; 10^{-10}$	$3.0 \; 10^3$

Collisional de-excitation of the fine structure transiton of C^+
can therefore be neglected in "typical" ISM situations, where
$n_e < 1$, $n_H < 300$, but might be important in the compressed layers
behind a strong radiative shock. The most important cases of
collisional de-excitation occur in HII regions; see [13] and M.
Perinotto's contribution. All this discussion becomes more
complicated when more than two levels are involved and/or the
ambient radiation field cannot be neglected.

2.3 "Neutral ISM" Regime

This regime corresponds to the parts of the ISM which are cool $(T < 10^4$ K) and shielded from ultraviolet photons with $h\nu > 13.6$ eV, but not from those with $h\nu < 13.6$ eV. The cooling function depends on ionization in two ways: i) through the ionization structure of the species which dominate the cooling (essentially C, Si, Fe, O) and ii) through the overall electron density that affects the importance of electron-induced inelastic collisions.

The first point can be settled easily over a wide range of conditions; see [14] and [15]. Elements having ionization potential $\chi > 13.6$ eV (H and O) are shielded and are mostly neutral; carbon is mostly ionized whenever both $n_H < 10^4$ and $N_H < 3 \cdot 10^{21}$ cm^{-2} (corresponding to $A_V \simeq 1.5$); these conditions can in fact be taken to define the diffuse interstellar medium.

A small fractional ionization of hydrogen, although not affecting the abundance of H^0 as a cooling agent, may affect the cooling rate because electrons are much more efficient than H

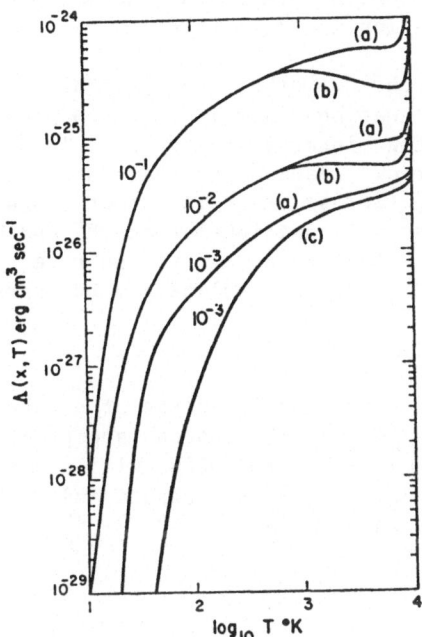

Fig 2.1 Cooling function in "neutral ISM" regime, for various electron concentrations. Curve (c) is obtained assuming that carbon is totally depleted ; this represents a bare minimum for cooling losses. Reproduced from Dalgarno and Mc Cray 1972 [16] by permission of the authors.

atoms for collisional excitation.   This contribution is not well
determined and depends on the assumed intensity of penetrating
radiation, soft X-rays or C.R. Electron excitation dominates for
$x = n_e/n_H > 10^{-3}$;   a   minimum   electron   density $n_e > n(C) \simeq$
$4 \cdot 10^{-4}$ n_H is present under the conditions outlined above.

The various processes relevant to that regime are extensively reviewed in [16].

2.4   "H^+ Region" Regime

In this regime, the gas is both heated and photoionized by
stellar UV photons with energies $h\nu > 13.6$ eV. Of the three
regimes that we discuss, it is the best understood; an "exact"
description is in principle possible.

The problem of ionization structure is the converse of what
it is for the "neutral ISM" case: the total electron density is
known immediately   ($n_e = n_H$, or slightly more depending on the
ionization of He); the ionization structure of the minor species
is determined by the balance between photoionization, recombination, and charge exchange, and changes with distance from the
star.   The main   cooling species are normally N^+, O^+, O^{++},   and
collisional ionization is negligible.

Contrary to the two other regimes that we discuss--neutral
and coronal--the heating rate is easy to compute. For each
photoionization, the internal energy pair is the kinetic energy
of the photoelectron.   Because a steady state ionization equilibrium must prevail (unless the point under consideration happens
to be within the thin transition between H^+ and H^0 regions), the
rate of photoionization is equal to the rate of recombination,
and the heating rate per unit volume can be expressed as:

$$h = n_e n_H \ \alpha \ \langle \varepsilon_i \rangle \qquad\qquad\qquad (2.7)$$

where $\langle \varepsilon_i \rangle$ is the mean kinetic energy of a photoelectron, which
depends on the shape of the photon spectrum, but not on its
intensity. Because each ionization will eventually be followed
by a recombination, we can anticipate and subtract the mean
energy of a recombining electron $\langle \varepsilon_r \rangle \simeq (3/2)kT$, to obtain the
net revenue after tax:

$$h_{net} = n_e n_H \ \alpha \ (\langle \varepsilon_i \rangle - \langle \varepsilon_r \rangle) \ . \qquad\qquad (2.8)$$

This must be balanced by the cooling rate:

$$\ell = n_e n_H \Lambda(T,A,I) \ .$$

Were it not for the fact that the cooling function Λ depends on the ionization structure, we would have an especially simple situation in which the equilibrium temperature would be independent both of the density of the gas and the intensity of the ionizing radiation. In fact, the following must be kept in mind:

i) The intensity of the radiation affects the ionization balance of minor species, and an increase of that intensity may in some cases correspond to a decrease of the resulting equilibrium temperature.

ii) The shape of the radiation spectrum changes (hardens) when going away from the source, because of the ν^{-3} dependence of the photoionization cross-section of H^0.

iii) Contrary to the case of the diffuse neutral ISM, collisional de-excitation can be important, so that the cooling rate can no longer always be expressed under the form of eq. (2.3).

iv) The cooling function depends on the abundances of the elements that give rise to the major cooling ions (N and O mostly).

However, leaving aside the most extreme cases of underabundance or the very hot central stars of some planetary nebulae, model calculations and observations show that the temperature of H^+ regions mostly remains in the range $7 \cdot 10^3 < T < 10^4$ K.

The dynamical study of H^+ regions is therefore considerably simplified by this approximation of an isothermal equation of state.

The reader will find in chapter 3 of [13] a more detailed analysis of the thermal balance in H^+ regions, tables for the essential atomic physics data, and a list of references.

2.5 Coronal Regime

Practically, this regime corresponds to high temperature and low density, and shall be more specific a little further. For the moment, consider a gas at temperature T, optically thin to all emitted photons, in a negligible ambient radiation field. Once again, the determination of the cooling rate hinges on the determination of the ionization structure (we assume that element abundances in gas phase are known). The only processes relevant to ionization equilibrium are collisional ionization and recombination (radiative and dielectronic). Charge exchange plays little role when the abundance of neutral species is very low, as is the case for collisional equilibrium at high temperatures. At equilibrium, the abundances of two consecutive ionization stages of element X obey:

$$n_e n_{x,s} i_{x,s}(T) = n_e n_{x,s+1} \alpha_{x,s+1}(T) \ . \tag{2.9}$$

The ionization structure is therefore independent of density, and is a function of temperature only. We can therefore rewrite equation 2.3; neglecting H-impact excitation, including bremsstrahlung, and assuming element abundances to be known:

$$-(dE/dt) = n_e n_H [\Lambda_e(T, I(T)) + b(T)]$$
$$= n_e n_H \Lambda(T) \ . \tag{2.10}$$

A few words about collisional ionization equilibrium. First note that this is not thermodynamic equilibrium: the abundance ratios derived from (2.9) are independent of electron density, contrary to what is the case for the Saha equation, the two processes in balance are not the reverse of each other, and the radiation field is way below the blackbody. Next, Figure (2.2), taken from [17] shows as a function of temperature the fractional abundances of successive ionization stages of oxygen (the H^0 –H^+ transition can be read off the same figure to an excellent approximation). Note that the transition from stage s to s+1 occurs when kT is still significantly lower than χ_s. Table 1 illustrates, in the case of oxygen, the comparison between the ionization potential of an ionic stage and the transition temperature, at which its abundance is equal to that of the next highest stage.

s	χ_s(eV)	kT_s(eV)
0	13.6	1.1
1	35	5
2	54	11
3	77	13
4	113	27
5	138	25
6	739	160
7	871	300

For typical values of photoionization cross-sections, using the observed values (or upper limits) of the diffuse soft X-ray flux, and assuming n_e to be tied to T by pressure equilibrium at $P/k = 4000$ k cm^{-3}, one can estimate that photoionization is negligible compared to collisional excitation for $T > 10^5$ K.

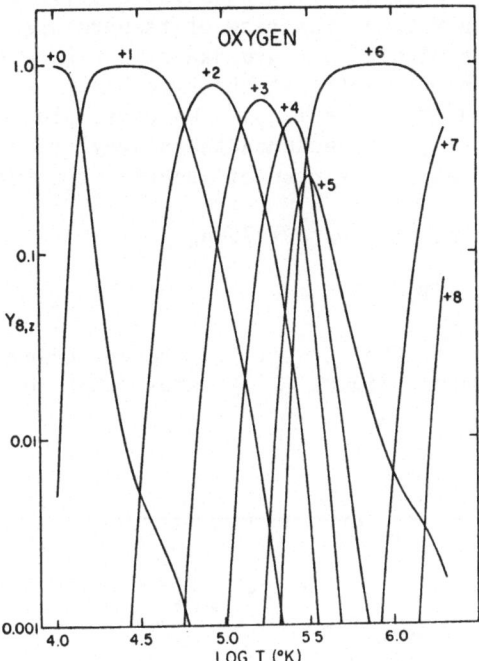

Fig 2.2 Fractional ionic abundances for oxygen in
collisional equilibrium. Reproduced from Shapiro and Moore
1976 [17], by permission of the authors.

Below 10^5 K, or for atypical density or EUV ambient flux, the
situation must be evaluated from scratch. The most recent
calculations of the cooling function are described in [18] and
[19]. Some less up-to-date references may be worth consulting
because the discussion of basic principles tends to be more
complete: [20], [21], [22], and chapter VI of [2] are somewhat
oriented to high density situations.

When is the assumption of steady state ionization equilibrium valid? To answer this, we must compare the timescale for
temperature change $(d(\log T)/dt)^{-1}$ with the timescale for ionization equilibrium. At or near equilibrium, each balance equation
of the form (2.9) has a characteristic rate (inverse timescale)
of approach to equilibrium

$$n_e r_{x,s} = (i_{x,s} + \alpha_{x,s+1}) n_e .$$

(2.11)

The assumption of collisional ionization equilibrium is valid if
that rate is larger than the rate of temperature change for all
ionic species of significant abundance. Figure (2.3) displays
$i_{x,s}$ for a number of species at the switchover temperature where
$n_{x,s} = n_{x,s+1}$ and $\alpha_{x,s+1} = i_{x,s}$. We have plotted on the same
figure the reduced (n_e dependence taken away) rate of cooling due
to radiative losses in the case of isobaric cooling:

$$n_e^{-1} r_{cool} = n_e^{-1} T^{-1} \; n_e^2 \Lambda(T)/5 k n_e$$

$$= \Lambda(T)/5kT \; . \tag{2.12}$$

In situations where the rate of temperature changes is governed
by radiative losses, figure (2.3) shows that the assumption of

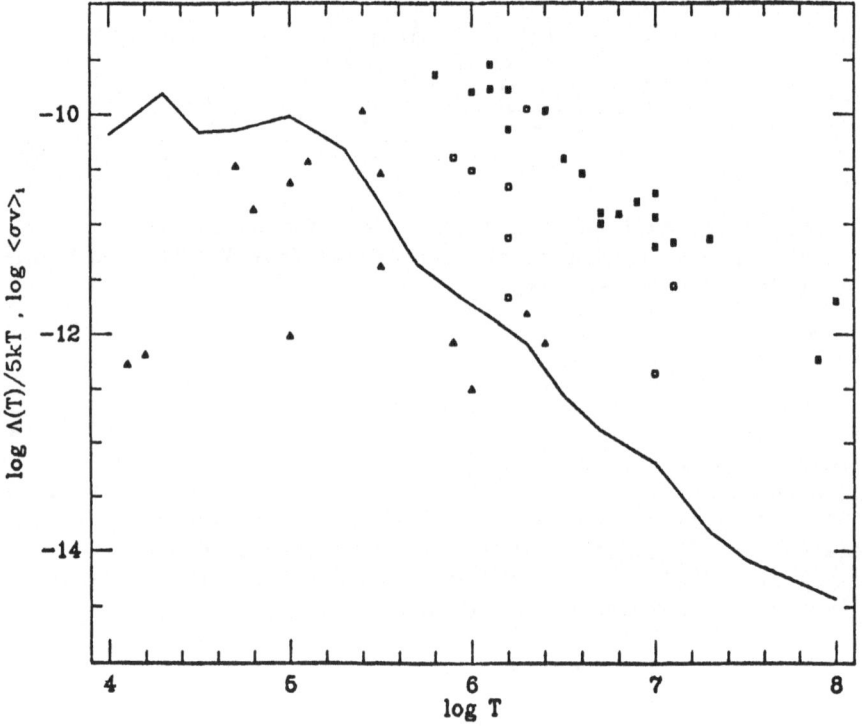

Fig 2.3 The rate of approach to collisional ionization
equilibrium for various ionic species. Triangles : carbon
and oxygen; open squares : Si^{+6} through Si^{+13}; filled
squares : Fe^{+6} through Fe^{+25}. The reduced rate of cooling
is plotted for comparison.

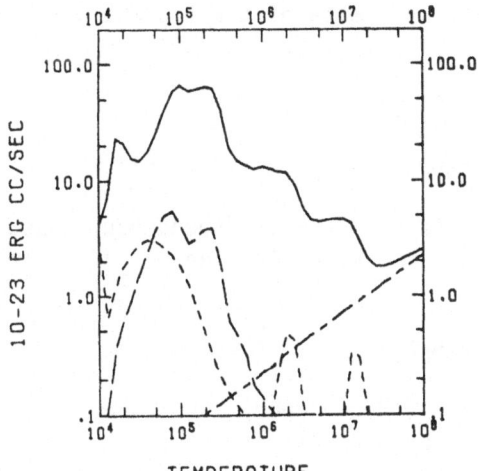

—Total cooling coefficient (————), forbidden line
cooling (- - - - - -), semiforbidden line cooling (— — —), and
bremsstrahlung (— — — —).

Fig 2.4 Cooling function in the coronal regime, with
equilibrium collisional ionization. Reproduced from
Raymond, Cox, and Smith 1976 [18], by permission of
the authors.

collisional ionization equilibrium is valid for log T > 6.5.   In
situations where the rate of temperature change is faster than
due to radiative losses (pdV expansion "cooling", conduction), a
specific comparison must be made.   The effect of a rate of
temperature change dT/dt on the cooling losses is that the
ionization structure lags behind the equilibrium;   for decreasing
temperature, the gas is over-ionized with respect to the equili-
birum at the instantaneous temperature, and the actual losses are
smaller than the equilibrium value $\Lambda(T)$, because, by and large,
excitation energies of collision-excited transitions tend to
increase from one ionization stage to the next.   We shall return
to this in section 3 when discussing the relaxation behind a
shock, where out-of-equilibrium initial conditions also play an
important role.

      Ionization rates can be found in [23], recombination rates
in [24] (errata in [25]), and a recent update in [26].

2.6 Dust Cooling

      Another energy sink is dust grains.   Ions or electrons
hitting dust grains tranfer to them part of their kinetic energy;
the grains achieve their energy balance by radiating in the

infrared.   A naïve estimate of the cooling due to that process
might be:

$$-(dE/dt)g \approx n_e n_g \sigma_g v_e \varepsilon_e$$

$$\approx n_e n_H \cdot (n_g \sigma_g / n_H)(k/m)^{1/2} T^{3/2} \quad . \qquad (2.13)$$

Dust cooling has an n^2 density dependence, just like atomic
collisional losses, to which it can be directly compared, by
defining:

$$-(dE/dt)_g = n_e n_H \Lambda_g(T) \quad . \qquad\qquad (2.14)$$

Because of the $T^{3/2}$ dependence, we expect dust cooling to
dominate at high temperatures.

    The actual computation of dust cooling is a lot more
complicated;   grain charge, grain transparency (for small grains
and high energy particles), and grain destruction by sputtering,
must all be taken into account.   This analysis is carried out in
[27], [28], [29]; typical results are illustrated in figure
(2.5).

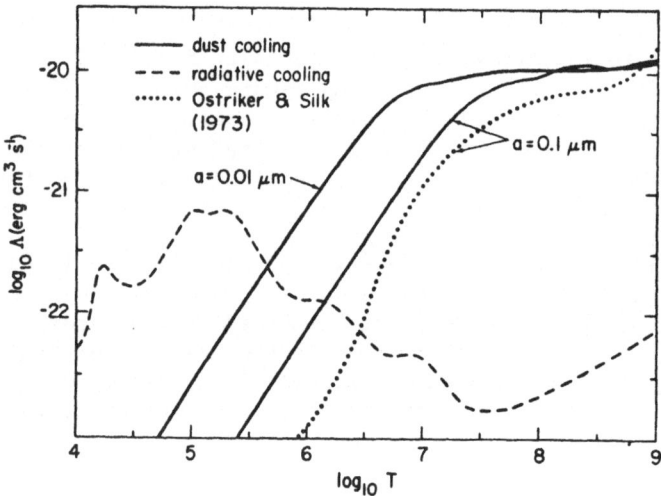

Fig 2.5 Dust cooling, expressed as equivalent cooling
function $\Lambda_g(T)$, for two values of grain radius.   For
comparison, gas radiative cooling [18], and a previous
calculation of grain cooling [27], are also plotted.
Reproduced from Dwek and Werner 1981 [29] by permission of
the authors.

Because dust cooling is limited by grain destruction, its importance can be assessed only in a specific situation. For supernova remnants, dust cooling is found to account for a significant, but not dominant, fraction of time-integrated radiative losses, between 10% and 35%, depending on the ambient density and the authors [30], [31]. For interstellar shocks, this fraction is estimated to be 10% for shock velocities $v_s > 200$ km s^{-1} [32]. For evaporating clouds, one can estimate from the results of [33] that dust cooling is not significant.

3. SHOCKS

In this section, we shall describe shocks from various points of view, from the most general to the most detailed, most "dirty," and most relevant to observational data. We shall first recall shock relations derived from conservation laws, then superficially address the problem of internal structure of adiabatic shocks, and finally study the radiative (and therefore observable) post-shock cooling region.

3.1 Adiabatic Shocks. General.

A shock is, ideally a finite discontinuity in the hydrodynamic variables of a fluid. Shocks can result from discontinuities in the initial conditions, or arise during the evolution of an initially continuous flow. For more details, consult paragraphs 93-95 of [1], or paragraph 9 of [2]. We shall assume that such a discontinuity exists and briefly recall some basic properties. By restricting oneself to a small region of space and time, and choosing a suitable frame of reference, one can assume the shock to be plane-parallel, stationary, steady-state, and the gas velocity to be perpendicular to the shock. One can therefore use results derived in section 1 for one dimensional time independant flows. Let ρ, u, ε be the density, velocity, and internal energy per unit mass, and let indices 1 and 2 denote the upstream and downstream sides of the shock. We set up a control volume having the shape of a pillbox or cylinder comprised between two plane surfaces parallel to the shock, and use equations (1.3). They reduce to:

$$[\rho u] = 0 \qquad\qquad\qquad (3.1a)$$

$$[p + \rho u^2] = 0 \qquad\qquad\qquad (3.1b)$$

$$[\varepsilon + P/\rho + u^2/2] = 0 \qquad\qquad\qquad (3.1c)$$

where square brackets [X] denote the jump of X from upstream to downstream. Given ρ_1, u_1, ε_1, the three equations (3.1) determine the downstream values. If we specialize to a perfect monatomic gas (γ = 5/3), the jump relations can be conveniently expressed as a function of the adiabatic Mach number $M = u/(\gamma p/\rho)^{1/2}$:

$$\rho_2/\rho_1 = u_1/u_2 = 4M_1^2/(3 + M_1^2) \qquad (3.2)$$

$$p_2/p_1 = 5M_1^2/4 - 1/4 \qquad (3.3)$$

$$M_2^2 = [(M_1^2 + 3)/(5M_1^2 - 1)] \qquad (3.4)$$

More general relations are found in paragraph 85 of [1]. Note that equations (3.1) do not discriminate between up-and down-stream quantities, and are invariant under time reversal ($u \rightarrow -u$). Note also, from (3.4), that $M_2 < 1$ whenever $M_1 > 1$ and vice versa. Therefore, to any solution of equations (3.1) involving a transition from super-to subsonic motion, there corresponds the reverse transition. However, the transition from super-to subsonic motion through a shock is an irreversible process involving an increase in the specific entropy; and the reverse process, although energetically allowed, has a vanishing probability of occurrence, just like the unmixing of, say, coffee and milk in a cup.

Weak shocks ($M_1 \simeq 1$) propagate at the adiabatic speed of sound, like all small disturbances.

Strong shocks correspond to the limit $M_1 \gg 1$; astrophysical shock with "interesting" observable effects are generally strong. For a strong adiabatic shock: i) the compression ratio $\rho_2/\rho_1 = 4$ ii) $M_2 = 5^{-1/2} \approx 0.45$ iii) ram pressure dominates the momentum flux upstream, but is only 1/4 of the total downstream iv) bulk kinetic energy dominates the energy flux upstream, but is only 1/16 of the total downstream v) the downstream temperature is given by: $kT_2 = (3/16)\mu u_1^2$ (3.5) where μ is the mean molecular mass of the gas vi) relative to the upstream gas, which often happens to be at rest in the frame where the problem is set up, the downstream gas moves at $3u_1/4$ When deriving equations (3.1) from equations (1.3) we actually made a careless leap and surreptitiously discarded a possible solution. Which one?

3.2 Adiabatic Shocks. Thermalization.

We have so far ignored the actual physics of the process by which kinetic energy is converted into thermal energy within the shock, relying on conservation laws alone. Consider the upstream particles impacting upon the downstream gas with relative velocity $3u_1/4$. Given binary collisions alone, they must slow down in one mean free path. For a neutral gas, the stopping distance $l_n = (n\sigma)^{-1}$, where $\sigma \approx 10^{-16}$ cm^2, and the shock thickness is smaller than any other relevant scale for a gas dynamics problem (e.g., the thickness of an ionization front $l_{if} \approx (n\sigma_i)^{-1}$, with $\sigma_i \approx 3.10^{-18}$ cm^2). For a fully ionized gas, the Coulomb mean free path increases like u^4; more specifically, the stopping distance of a pre-shock proton in the post-shock gas is:

$$\lambda_c = 0.1 \text{ pc } (u_1/1000 \text{ km s}^{-1})^4 n_1^{-1} \qquad (3.6)$$

which can be comparable to other flow length scales for a fast shock moving into a low density medium. However, binary particle collisions are just one possible mechanism for the thermalization of bulk kinetic energy, and the stopping distances computed above are only upper limits to the shock thickness. Other possible mechanisms are collective instabilities or photon diffusion (for optically thick shocks dominated by radiation pressure). Indeed, shocks in the interplanetary plasma are observed to have a thickness much smaller than the Coulomb stopping length. One can probably assume that the same is true for interstellar shocks (see, however [32]). One may also ask over what distance (or in how much time) electrons and ions achieve a common temperature behind a strong shock in a fully ionized gas. Although the lack of equipartition does not affect the dynamics (because $p = (\gamma-1)\rho\varepsilon$ holds even if $T_i \neq T_e$), it does affect spectroscopic diagnostics, which, except for linewidths, depend on T_e alone. Using the equipartition time given in chapter 5 of [33], we find the equipartition length behind a strong shock:

$$\lambda_{ei} \approx 0.4 \text{ pc } (u_1/1000 \text{ km s}^{-1})^4 n_1^{-1} \qquad (3.7)$$

roughly equal to the stopping distance found above (the expected $(m_p/m_e)^{1/2}$ factor is partly cancelled by the fact that pre-shock protons are suprathermal relative to post-shock gas). In situ observations show that $T_e \neq T_i$ behind interplanetary shocks; however, eq 3.7 gives only an upper limit to the equipartition length, collective processes might heat up the electrons faster.

For partially ionized gas, charge exchange plays a role in the relaxation layer to couple neutral and ionized particles (see also [37]).

3.3 Magnetic Shock. Jump Relations.

We now briefly examine the case of shock with a magnetic field. When the field is parallel to u_1, equations (3.1) are not modified. When it is perpendicular, the equations must be modified and supplemented as follows:

$$[\rho u] = 0 \qquad (3.8a)$$

$$[p + \rho u^2 + B^2/8\pi] = 0 \qquad (3.8b)$$

$$[\varepsilon + P/\rho + u^2/2 + B^2/4\pi\rho] = 0 \qquad\qquad (3.8c)$$

$$[Bu] = 0 \qquad\qquad (3.8d)$$

Equations 3.8 now include magnetic energy and pressure terms, and an extra equation expresses the conservation of magnetic flux in a conducting plasma. The solution for the downstream conditions is slightly more complicated than in the unmagnetized case; we only mention that, all other upstream parameters unchanged, the presence of a magnetic field magnetic field decreases the compression across the shock, and that the minimum upstream Mach number (weak shock limit) is no longer 1 but :

$$(M_{1,min})^2 = 1 + 6\alpha/5$$

where $\alpha = B_1^2/8\pi p_1$ is the upstream ratio of magnetic to gas pressure. Note that a jump in B implies that a surface current is flowing in the shock front. For weakly ionized gases and/or high values of B, the associated dissipation can entirely mediate a continuous ("C-type") shock; see [32a].

The jump conditions for an arbitrary orientation of the magnetic field can be found in [33].

3.4. Radiative Shocks. General.

Behind a shock, both the density and the temperature increase. Except for the fastest shocks ($u_1 > 1000$ km s^{-1}), the cooling timescale $t_{cool} = kT/n\Lambda$ will be shorter behind than ahead of the shock. It is, however, long enough that a clear distribution can be maintained between the shock proper (relaxation to Maxwell-Boltzmann, energy conserved) and the cooling region where radiative losses take place. The structure made up of an adiabatic shock and the following cooling region is called a radiative shock. Note that an interstellar shock is not adiabatic or radiative per se, the distinction can only be made by comparing the cooling length with other typical scales for the problem at hand. If the cooling length is short enough compared to the other scales of the problem, the structure of the cooling region can be studied under the assumption of steady-state. We can rewrite the conserved momentum flux, introducing the compression ratio $k = \rho/\rho_1$:

$$\Pi = (2/3)k\rho_1\varepsilon + k^{-1}\rho_1 u_1^2 + k^2 B_1^2/8\pi \qquad\qquad (3.9)$$

From now on, we shall assume that we deal only with <u>strong shocks,</u> i.e. a strong adiabatic shock followed by a cooling region. Assume for the moment that magnetic pressure is small, which is true immediately behind the adiabatic shock if the upstream velocity is large compared to the Alfven velocity (i.e. $\rho_1 u_1^2 \gg B_1^2/8\pi$). As radiative losses rob the internal energy ε, k must respond to maintain constant the r.h.s. of (3.9). Since the first term dominates over the second, the change of k must be an increase. As ε continues to descrease, the first term becomes more dominant. Since gas pressure is 3/4 Π just behind the shock, and increases downstream, a fairly good approximation is to assume that the cooling region is isobaric. As the post-shock gas cools down and gets compressed, it becomes essentially comoving with the shock front (see eq 3.8a). This simplifies observational diagnostic.

This process of cooling and compression can be terminated in either of two ways, whichever comes first.

a) The temperature drops low enough that radiative losses become negligible and/or are compensated by a heating term. Let the index 3 denote the asymptotic values,

$$\Pi = \rho_1 u_1^2 = P_3 = (P_3/\rho_3)k_3\rho_1 \qquad (3.10)$$

where we have left aside the ram pressure term in the final "3" state. If the final temperature is known, the compression factor can be deduced simply from:

$$k = (u_1/c_3)^2 \qquad (3.11)$$

where c is the isothermal speed of sound, defined by:

$$c^2 = P/\rho = kT/\mu \qquad (3.12)$$

The complete structure is called an isothermal shock, although the pre-shock value of the temperature need not be equal to the final one, which is the only one that matters (provided the shock is strong). Situations where the final temperature is known include:
i) photoionized gas (T $\simeq 10^6$K, c $\simeq$ 10 km s^{-1})
ii) neutral gas, if thermal equilibrium can be deduced from the known post-shock pressure and some heating mechanisms (see Section 8).

b) The magnetic pressure contribution, growing as k^2, may eventually dominate the r.h.s. of (3.9). Further increase of the density is then prevented, although the gas may continue to cool down. Equation (3.11) is no longer valid, but one can relate the shock velocity to the total (gas and magnetic) pressure driving the shock from downstream, whatever its nature:

$$\rho_1 u_1{}^2 \quad = \quad P_{down,tot} \qquad\qquad (3.12)$$

where the error due to the neglect of downstream ram
pressure is at most 25%.

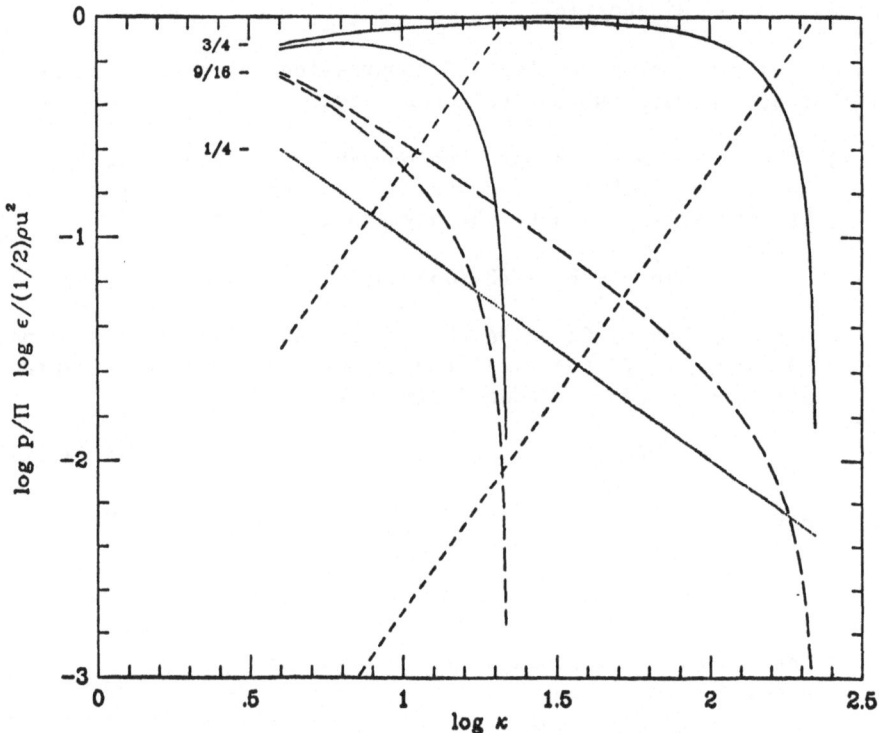

Fig 3.1 Evolution of the momentum budget behind a
strong radiative shock with a transverse magnetic field, as
a function of the compression ratio κ. Full line : gas
pressure; dotted line : ram pressure; short dash :
magnetic pressure. Also shown, as long dashed line, the
specific internal energy. Two cases were computed for
different values of the pre-shock ratio $\delta = B^2/(8\pi\rho_1 u_1{}^2)$:
$2\ 10^{-5}$ and $2\ 10^{-3}$, reaching maximum compression $\kappa = 225$ and
22.5 respectively. Representative values might be $n_1 = 1$
cm^{-3}, $u_1 = 100$ km s^{-1}, and $B_1 = 0.3$ μG and 3 μG
respectively.

Figure (3.1) shows the variation of the three parts of the momentum flux in eq. (3.9) behind a strong shock (normalized to the total momentum flux) as a function of compression ratio k. Also shown is specific internal energy ε (normalized to upstream specific internal energy $u_1^2/2$). Two cases are shown, for two values of: $\delta = B_1^2/8\pi\rho_1 u_1^2$, the upstream ratio of magnetic to ram pressure; $\delta = 2.10^{-3}$ and $\delta = 2.10^{-5}$. A possible set of physical values might be:

$n_1 = 10 \text{ cm}^{-3}$

$u_1 = 100 \text{ km s}^{-1}$

$B_1 = 1\mu G \ (\delta = 2.10^{-5}), \ 10\mu G \ (\delta = 2.10^{-3})$
Temperature (internal energy) scale factor $T^* = 2.42 \ 10^5$ K
Post-shock temperature: $T_2 = 1.36 \ 10^5$ K $= (9/16)T^*$

The pre-shock gas is assumed to be fully ionized, He abundance 10% by number.

If one defines a point in the cooling region by its temperature $T < T_2$; figure (3.1) allows one to find the density and the importance of magnetic pressure at that point.

3.5 Radiative shocks. Models.

The discussion of the previous subsection allowed us to find the evolution of internal energy and various pressure terms with density as a parameter. If we want to determine these as a function of distance from the shock, we must equate the divergence of the energy flux to the radiative sink term:

$$\frac{d}{dr}\left[\rho_1 u_1\left((5/3)\varepsilon(r) + k^2 B_1^2/4\pi\right)\right] = -k^2 n_1^2 (n_e/n)\Lambda \qquad (3.13)$$

At first sight, equations (3.13) and (3.9) allow us to solve for ε and k as a function of distance r. However, for most radiative shocks of interest, the cooling gas is out of collisional equilibrium, so that, as we stressed in section 2, Λ is not a function of ε alone, but also of the whole set of instantaneous ionic abundances. Their variation must be integrated through a set of coupled differential equations, and initial values must be provided. Besides, within a radiative shock, the UV-EUV-SXR radiation field is not the average interstellar radiation field, and photoionization cannot be neglected.

Sucessive publication of numerical models for radiative shocks have been increasingly refined. All incorporate time-dependant equations for the ionized fractions in the post-shock gas, including photoionization by shock radiation. In [33],

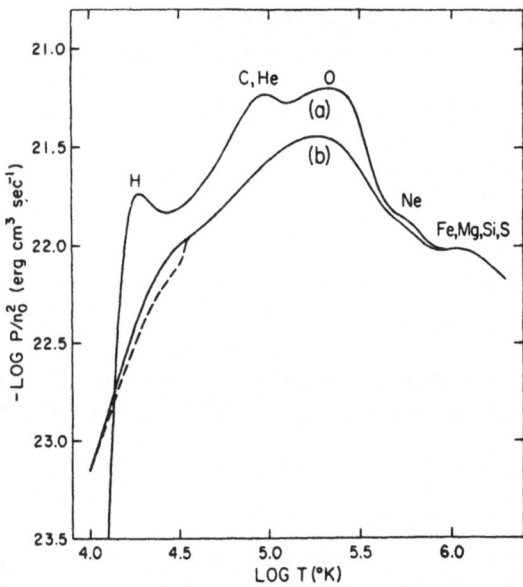

Fig. 3.2 Cooling function $\Lambda = P_{rad} / n^2$ for a) steady-state
ionization equilibrium; b) time-dependant isochoric cooling,
initially in ionization equilibrium at $T = 10^6$ K. The
difference between the two cases is significant but not
dramatic. Reproduced from Shapiro and Moore 1976 [17], by
permission of the authors.

hydrogen is assumed to be fully ionized in the pre-shock gas, and
other elements are assumed to reach instantaneously ionization
equilibrium at the post-shock temperature.   In [34] and [35],
educated guesses are made concerning the pre-shock ionization
structure that might result from EUV radiation emitted by the
shock itself.   In general, this does not coincide with collisional equilibrium at the post-shock temperature, and the ensuing
adjustment taxes the internal energy by a far larger amount than
just the difference in ionization energies.

To further examine this point, we first resort to the
isochoric cooling calculations of [36] and [17].   Fig. (3.2)
compares the time-dependant cooling rate for a gas cooling at
constant volume, to that computed assuming equilibrium ionization
at each temperature.   Constant pressure cooling would be more
relevant to radiative shocks, but the difference is not essen-

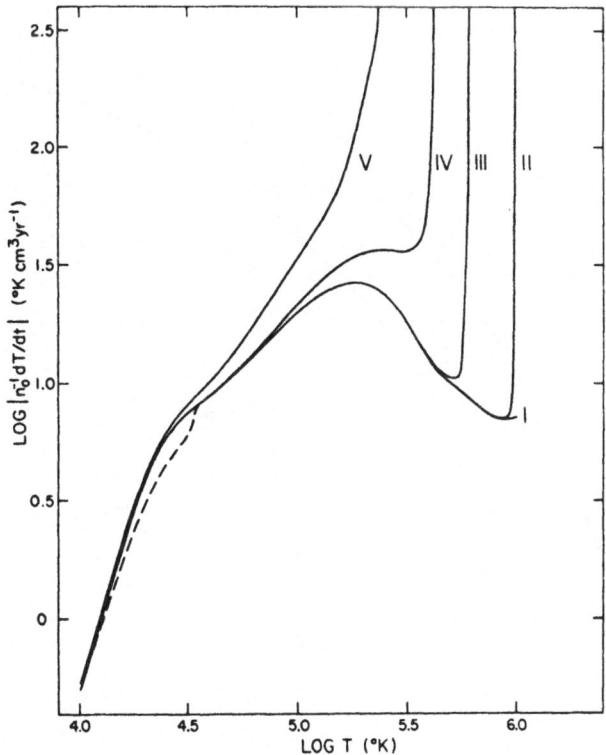

Fig. 3.3 Cooling rate versus temperature for gas initially at 10^6 K, but with an ionization structure corresponding to various initial conditions. I : collisional equilibrium (CE) at 10^6 K; II : photoionized by 40 eV photons; III : CE at $1.56 \cdot 10^4$ K (H 50% neutral); IV : CE at $1.42 \cdot 10^4$ K (H 75% neutral); V : CE at 10^4 K (H 100% neutral)

tial, because cooling, ionization, and recombination times all scale as ρ^{-1}. The gas is assumed to be initially at T =10^6 K in collisional equilibrium. Time-dependant ionization for a cooling gas is seen to result in a significant but not dramatic decrease of radiative losses. If the gas is set up initially at T = 10^6 k, but under-ionized for that temperature, the radiative losses are initially larger than the steady-state value by orders of magnitude. Once the ionization structure has reached a rough agreement with the temperature, the cooling curve follows a common track (see figure 3.3) that coincides with the time-dependant cooling curve of figure 3.2; the initial conditions

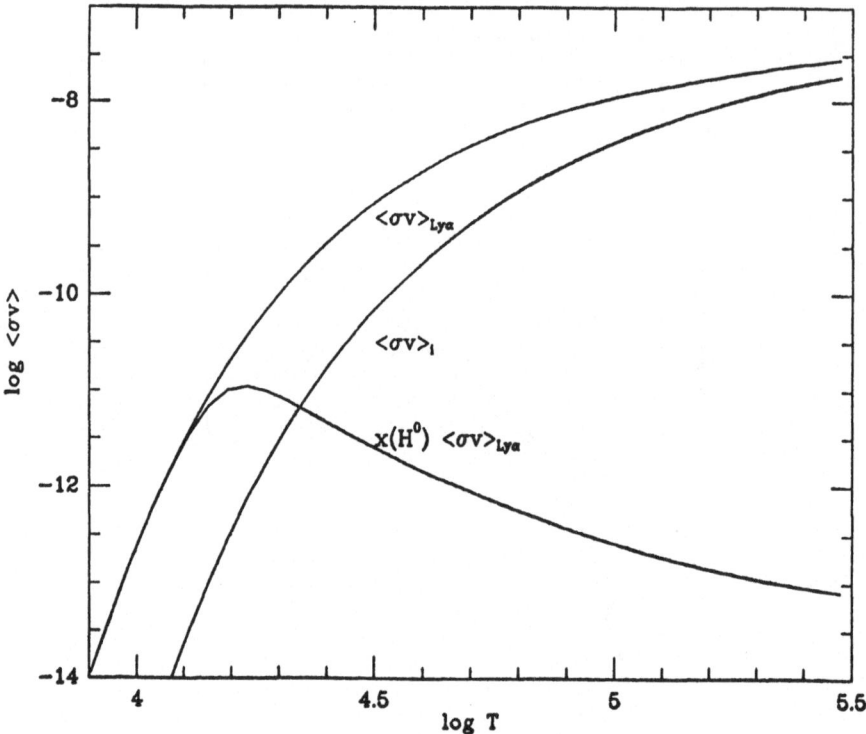

Fig. 3.4 Rate of collisional excitation of hydrogen (1s-2p) per neutral H $\langle\sigma v\rangle_{Ly\alpha}$; effective rate $x(H^0)\langle\sigma v\rangle_{Ly\alpha}$; and rate of collisional ionization $\langle\sigma v\rangle_i$.

have been forgotten. Curve V of figure 3.3 corresponds to a case where the gas was initially neutral at 10^6k; its internal energy is initially $1.52\ 10^{-10}$ erg/H atom, and it loses more than 90% of that amount before approaching ionization equilibrium, out of which only $0.26\ 10^{-10}$ erg/H atom has been invested in ionization energy. The large enhancement of radiative losses can be understood with the help of figure 3.4. The rate of collisional excitation of Lyα rises steeply with temperature, and the effective rate decreases above $2\ 10^4$ K only because the neutral fraction $n(H°)$ does. If, however, the ionization structure is out of equilibrium, Lyα losses can be enhanced by orders of magnitude. For each H° that has to be collisionally ionized (cost 13.6 eV), between 3 and 10 Lyα photons (cost 10.2 eV) are emitted. The pre-shock ionized fraction is therefore an important parameter in model shock calculations.

This point has been addressed in [37], where the photo-ionization of the pre-shock gas has been taken into account, giving self consistent pre-shock ionized fractions.

To summarize, a radiative shock can include up to five regions:

i)   photoionized precursor $1 \simeq 3.10^{17} \; n_1^{-1}$ cm

ii) shock proper, relaxation to kinetic equilibrium, length determined by the dominant dissipative process (ion viscosity, finite conductivity, plasma instabilities).

iii) relaxation towards collisional ionization equilibrium, $1 \simeq 3.10^{13} \; u_{100}^{1/2} \; \exp(1.2/u_{100}^2) n_1^{-1}$ cm

iv) cooling at constant pressure, moderate deviations from collisional ionization equilibrium

v)   high density, low temperature region where photoionization, magnetic pressure, and collisional de-excitation may become important.

The published numerical models probably give an adequate description of radiative shocks in the diffuse ISM. However, since parameter space is at least three-dimensional, the inter-pretation of some line intensities (those originating in region (v) defined above, in particular) cannot be made, from published data alone. The interested reader is invited to read the discussion sections in [35] and [37]. A profitable exercise might be to qualitatively analyze the trends in the diagnostic diagrams of [36].

4. IONIZATION FRONTS

In this section, ionization fronts are first analyzed and classifed using conservation laws. The apparent lack or excess of solution for some values of the parameters is resolved by using a stability analysis and constraints on the possible realizations of the internal structure.

4.1 Conservation Laws and Classification

A region of ionized gas develops when a flux of Ly-c photons is incident upon neutral gas. For densities of neutral gas commonly found in the ISM ($n_H \gtrsim 1$ cm^{-3}), and photon energies not too much above 13.6 eV, the mean free path of a photon in the neutral gas $\lambda \simeq 3.10^{17} \; n_H^{-1}$ cm is small compared with typical ISM scale lengths. The transition between neutral and ionized gas is therefore sharp, and can be considered as a discontinuity

We assume (recall section 2.4) that the photoionized gas is adequately described by an isothermal ($T = T_2$) equation of state. We shall assume that the neutral gas, subject to radiative cooling and some unspecified heating mechanism, achieves thermal equilibrium at a constant temperature T_1, with a short equilibrium timescale. We shall make use of the isothermal sound speed $c = (p/\rho)^{1/2}$ and the isothermal Mach number $m = u/c$. We assume that the initial c_1 is less than the final value c_2; this is always true in practice.

In the frame of the ionization front, the conserved mass and momentum fluxes are:

$$J = \rho_1 u_1 = \rho_2 u_2 \qquad\qquad\qquad (4.1)$$

$$\Pi = p_1 + \rho_1 u_1^2 = p_2 + \rho_2 u_2^2 \qquad\qquad (4.2)$$

We have 2×3 variables describing the up and downstream gas. Given the upstream density, the two isothermal equations of state, and the two conservation laws, there rmains one degree of freedom. We can equivalently specify J, u_2, or m_2. As for shocks, a convenient parameter is $k = \rho_2/\rho_1$. After some elementary algebra, one obtains for k a quadratic equation which has:
  - two solutions for $m_1 < m_{1,CD} = c_1/2c_2$ ($u_1 < c_1^2/2c_2$); these are called D-type
  - no solution for $m_{1,CD} < m_1 < m_{1,CR}$
  - two solutions for $m_1 > m_{1,CR} = 2c_2/c_1$ ($u_1 > 2c_2$); these are called R-type.

We introduce at this point one extra assumption for the sake of comprehension: that recombinations are negligible. This allows one to equate the particle flux J/μ wth the incident photon flux Φ, and to think of the J-front as being directly "driven" by Φ. This assumption is in fact not exact. This being said, for a given photon flux, the upstream velocity will be large or small according to whether the upstream gas is dense or rarefied. Now, in each of the "allowed" (D and R) regions for the upstream Mach number, there are two branches, called weak and strong; at a given value of m_1, the wD solution, say, has a smaller change in density (smaller $|Ln(k)|$) than the sD solution. At the limiting values of m_1, i.e. $m_{1,CD}$ (resp. $m_{1,CR}$) for the D region (resp. R region), the two branches join at the critical cD (resp. cR) solution. The preceding discussion is fairly classical and can be found in [3], [4] or [13]. The classification of I-fronts was first established in [38]; even though a different energy condition is used, this paper is still well worth reading. Table 4.1 summarizes the properties of the various branches. Note that for the critical solution, the downstream flow is just sonic. Anticipating on what follows, focus attention on the weak branches. Consider first the asymptotic behavior. One sees easily that the extreme-wR front moves at high velocity and leaves the density undisturbed; it can be described as "flash" ionization. The extreme-wD front moves at vanishing velocity and makes a constant pressure transition; a quasi static situation.

	D	cD	m_1	cR	R	∞	
	0	$\dfrac{c_1}{2c_2}$		$\dfrac{2c_2}{c_1}$			
wD	$\dfrac{c_2 m_1}{c_1}$ <1	1	m_2	1	>1	$\dfrac{c_1 m_1}{c_2}$	wR
sD	$\dfrac{c_1}{c_2 m_1}$ >1				<1	$\dfrac{c_2}{c_1 m_1}$	sR
wD	$\dfrac{c_1^{\,2}}{c_2^{\,2}}$ $>\kappa_{cr}$	$\dfrac{c_1^{\,2}}{2c_2^{\,2}}$	κ	2	<2	1	wR
sD	$m_1^{\,2}$ $<\kappa_{cr}$				>2	$\dfrac{m_1^{\,2} c_1^{\,2}}{c_2^{\,2}}$	sR

Table 4.1 A summary of some properties of the four types of I-fronts. The input isothermal Mach number ranges from 0 to ∞. Below are given characteristic values (for critical condition) and asymptotic behavior (for extreme types) of the exhaust Mach number m_2 and compression κ.

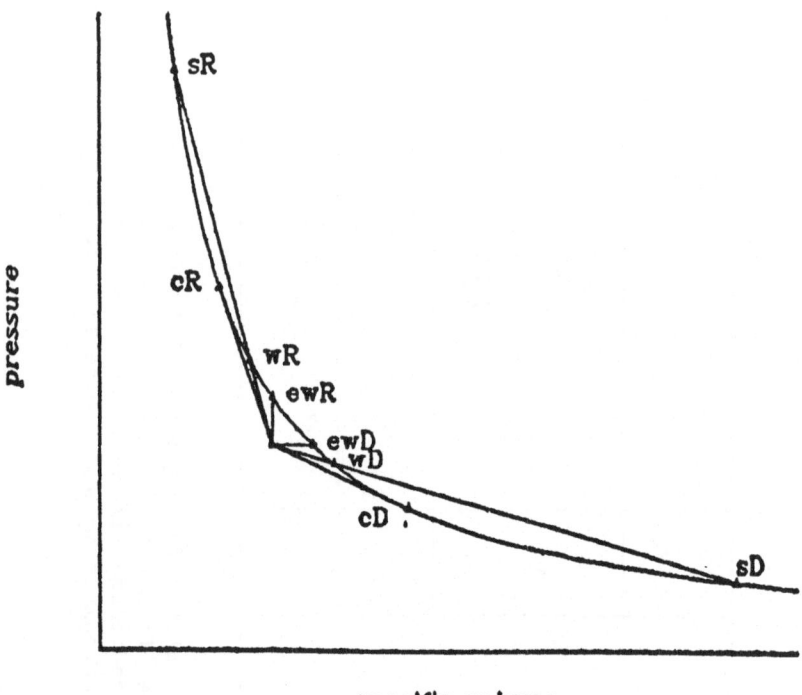

Fig. 4.1 The various types of I-fronts in the Hugoniot
diagram. Note that the "forbidden range" of values of the
mass flux corresponds to the lines through the initial state
that do not intersect the ionized gas isotherm; and not to
the quadrant between the extreme weak D and the extreme weak
R fronts; the latter are even more strongly forbidden (if
one may say so) because $J^2 < 0$!

The constant density (resp. constant pressure) is good within a
factor of 2 on the whole wR (resp. wD) branch.
    A better insight into the classification of I-fronts can be
gained by considering a graphical representation in the specific
volume ($V = 1/\rho$) - pressure diagram, called the Hugoniot diagram.
It derives its usefulness from the following properties, which
the reader can easily derive for himself:

i) if a jump transition connects two points (V_1,p_1), (V_2,p_2), the mass flux through that transition satisfies:

$$J^2 = -\Delta p/\Delta V = -(p_2-p_1)/(V_2-V_1) \qquad (4.3)$$

ii) the super- or subsonic nature of the flow ahead of or behind a jump transition can be determined by comparing the slope $\Delta p/\Delta V$ of the transition line with the slope of the equation of state dp/dV in the initial or final state. Given the isotherm $pV = c_2^2$ for the ionized gas and an initial state below $(c_1 < c_2)$ that isotherm, one can easily derive the qualitative features of the classification. In particular, the critical I-fronts are represented by the tangents from the initial state to the final isotherm.

4.2 "Forbidden" Range. Hands-on Approach.

Pending further analysis we assume that only weak (R or D) I-fronts exist. This can be deduced from a plausible assumption about the internal structure of I-fronts: that $c^2 = kT/\mu$ rises in a monotonic way from its initial to its final value. This excludes continuous transitions from the region of the Hugoniot diagram above the $pV = c_2^2$ isotherm. Strong-D fronts are thus excluded, but not strong R fronts with embedded shocks.

Consider a wR I-front moving into a uniform upstream gas, and let the photon flux Φ decrease continuously. The particle flux through the front J, and the slope of the transition line in the Hugoniot diagram (see fig. 4.2a) also decrease until the transition line is tangent to the final isotherm and a cR condition is reached. We have apparently reached an impasse. But just at this point, an isothermal shock can be inserted, that modifies the upstream conditions in such a way that a cD type front can match the same downstream conditions. If the photon flux continues to decrease, the shock will detach ahead of the D-type I-front, the precise nature of which (critical or weak) depends on the downstream boundary conditions (figure 4.2b). Replacing a wR front by a shock followed by a D front, before the cR condition was reached, would have required a sD front, which is excluded by our assumptions.

4.3 Stability Analysis

Here we do not impose on the possible I-fronts any constraint other than the conservation laws, and we do a local stability analysis of the propagation of an I-front. We follow a method similar to that used in paragraph 84 of [1] to study the stability of shock waves. The nature of constraints and of possible perturbations is different, however, because we have

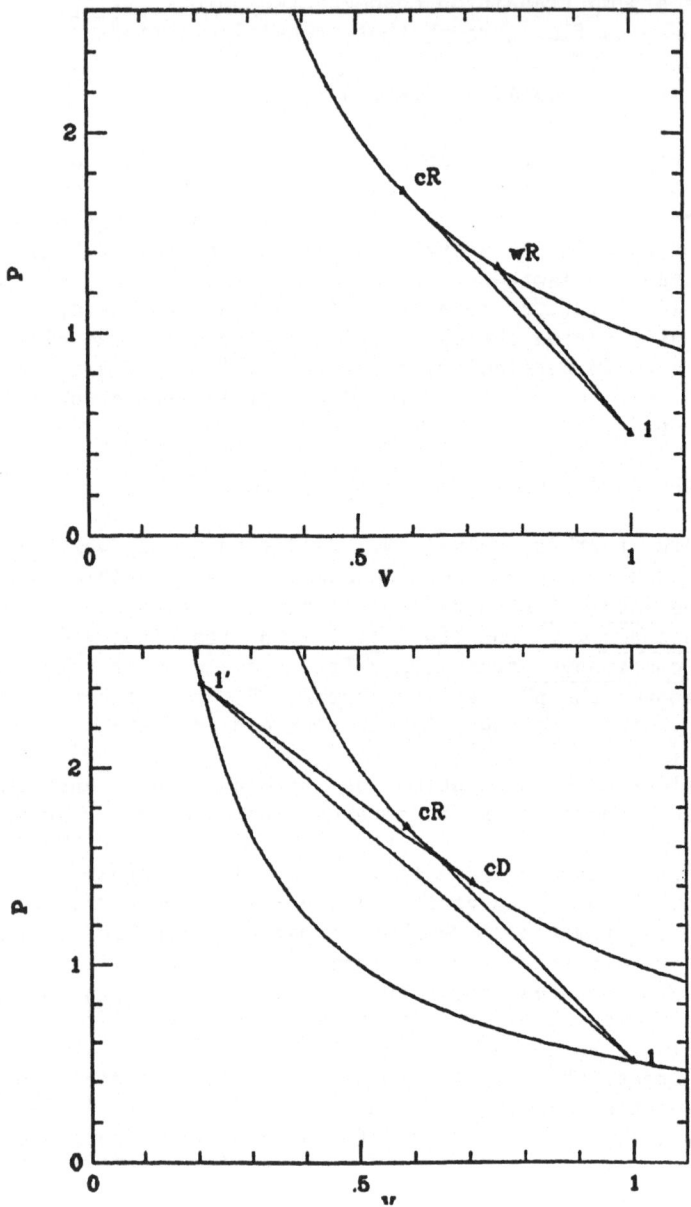

Fig 4.2a and 4.2b . Illustrating the smooth transition from
wR to shock+wD. See text.

assumed isothermal equations of state. Let (ρ_1, w_1), (ρ_2, w_2) define the state of the gas ahead and behind a front moving at velocity w_f. To determine whether perturbations away from that state are possible, we enumerate constraints and degrees of freedom.

Constraints.

$$J = \rho_1(w_1 - w_f)$$

$$J = \rho_2(w_2 - w_f)$$

$$\rho_1[c_1^2 + (w_1 - w_f)^2] = \rho_2[c_2^2 + (w_2 - w_f)^2]$$

The photon flux is unperturbed because we assume that recombinations are negligible.

Degrees of freedom. One is the perturbation of w_f. The fluid variables, ρ_1 and w_1, for instance, cannot be counted as degrees of freedom without qualification, because their perturbations must satisfy the flow equations. The two Riemann invariants, however, are decoupled and propagate independently of each other (see end of section 1), at the speed of sound (isothermal in our case) relative to the fluid. Take the case of a wR front, which is supersonic both upstream and downstream (figure 4.3). Both upstream characteristics run <u>into</u> the front; their Riemann invariants are determined by boundary conditions and cannot be perturbed. Downstream, both characteristics propagate <u>from</u> the front, and both their invariants can be perturbed. We leave it to the reader to complete the analysis for the three other types of fronts, leading to the results shown in Table 4.2.

Table 4.2

Front Type	wR	sR	wD	sD
Upstream motion	super	super	sub	sub
Upstream deg. freed.	0	0	1	1
Downstream motion	super	sub	sub	super
Downstream deg. freed.	2	1	1	2
Total deg. freed.	3	2	3	4

The total number of degrees of freedom includes the front velocity w_f. The conclusion is that wR and wD fronts are stable (there are just enough conditions to constrain the perturbations to zero), sR fronts are overconstrained (they cannot in general occur for arbitrary boundary conditions), and sD fronts are underconstrained (arbitrary perturbations, compatible with

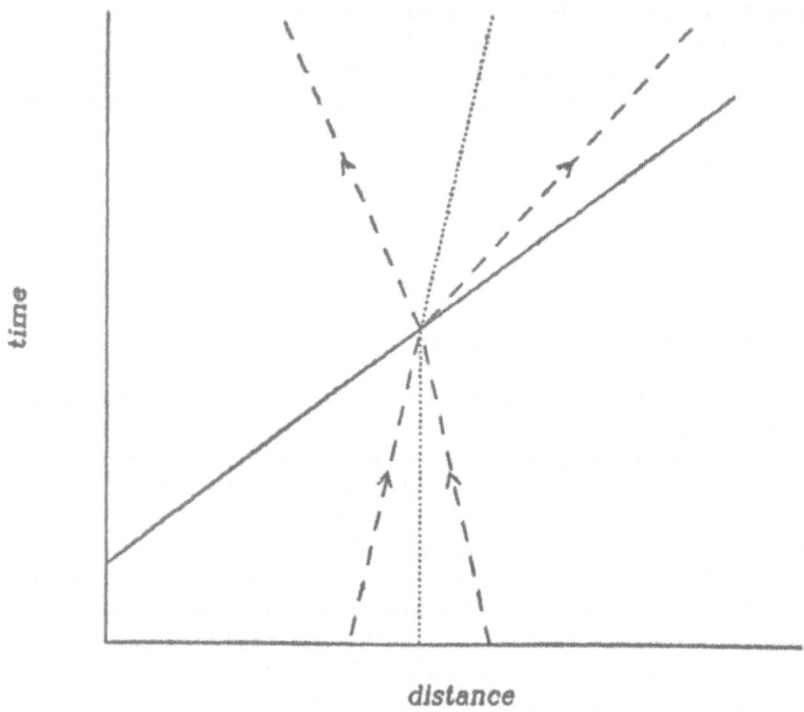

Fig 4.3 Space time diagram showing the relations between
characteristics (dashed) and a wR front (full line) which is
supersonic both upstream and downstream.   Also shown (dotted) the trajectory of a gas particle, sometimes called a C^0
characteristic.   The reader is invited to draw similar
diagrams for the other cases.

boundary conditions, are possible).   Note that shock fronts,
which, like sR I-fronts, make a transition from super to subsonic
motion, are not overconstrained because they can freely choose
their mass flux, while an I-front must obey a boundary condition
relative to the photon flux.

4.4 Actual Fronts. Uniqueness of Solution.

A general problem of interest can be stated thus: given
suitable boundary conditions, an incident photon flux, and a set
of allowed ionization fronts, what is the resulting flow, and is
it unique.   We gave an illustrative solution in 4.2, which was
incomplete because hydrodynamical boundary conditions were left
unspecified, and schematic both because we excluded sD fronts

through a plausible but unproven assumption, and because we neglected recombinations.

The first point can be readily cleared up. The solution we gave is valid for the upstream gas at rest, and a fixed downstream boundary pressure $p_0 < (c_2/c_1)^2 p_1$. The ionization front is followed by a region of uniform motion and by a rarefaction wave. More generally, for arbitrary photon flux and imposed downstream pressure, retaining the two simplifying assumptions about recombinations and sD fronts, it is a simple (but too lengthy to be described here) exercise to show that there is always a unique solution comprising an I-front and one or more of the following: shock, rarefaction wave, uniform motion. In these solutions, sR fronts occur only for isolated sets of boundary conditions as a limiting case of wR-shock or shock-wD superpositions.

The second point, of properly specifying which I-fronts, among those allowed by conservation laws, do actually occur, was first addressed in [39], and more recently in [40] and [41]. A welcome result is that, when sD fronts occur, they do for unique values of the upstream (and downstream) Mach numbers. The undeterminacy of sD fronts revealed by the stability analysis in 4.3 is thus resolved by the constraints imposed by the internal structure of actual I-fronts. Whether sD fronts do occur at all depends on the spectrum of the incident ionizing radiation, the temperature of the upstream gas, and the atomic physics (cooling, etc.) used. The internal structure integrations of [40] and [41] show that, for an upstream gas of normal composition, reasonable values of the upstream temperature ($T_1 < 1000$ k), and ionizing stars with temperatures $T_* < 40,000$ k, no sD fronts occur. Our simple argument is thus legitimized. As for sR fronts, they can always be constructed by inserting a shock close to or within a wR front, but our earlier conclusion from stability analysis still applies: they will occur only for particular, isolated sets of boundary conditions.

Recombinations are, of course, included in the front structure integrations cited above. They also occur in the downstream flow, and the actual boundary condition is not the photon flux at the front (as we assumed for the sake of simplicity) but at some point further downstream. This can lead to global flow instabilities, caused by the lag of the neutral fraction $x(H^0)$ on its equilibrium value, and having a characteristic timescale of the order of the recombination time $t_r = (\alpha n_e)^{-1}$; such instabilities have been studied in [42], [43], [44].

The problem of uniqueness of a flow involving an ionization front has not been solved in general. Solutions for interesting cases lending themselves to an analytical approach are given in

[45], [46], [40]. In a different approach, instead of considering I-fronts as discontinuities, their internal structure is
incorporated in the time dependent hydrodynamical equations for
the general flow, which are solved by finite difference techniques [47]. Other examples will be given in sections 6 and 7.

5. HEAT CONDUCTION

In a fully ionized plasma, the mean free path of a thermal
($m_e v^2 = 3kT$) electron is:

$$\lambda_e \simeq 4 \times 10^3 \ T^2 \ n^{-1} \ cm \qquad\qquad (5.1)$$

and, when λ_e is small compared with $\lambda_T = (\nabla Log T)^{-1}$, one expects
the heat flux to be on the order of:

$$F \simeq n_e \langle v_e \rangle \lambda_e \nabla (kT)$$

$$\simeq \kappa \nabla T \qquad\qquad (5.2)$$

where κ is independent of n and grows like $T^{5/2}$. Detailed
calculations [48] give the result:

$$\kappa = 0.6 \times 10^{-6} \ T^{5/2} \ erg \ cm^{-1} \ s^{-1} \ K^{-1} \qquad\qquad (5.3)$$

which is valid for a plasma containing 10% He by number, and
assuming Log $\Lambda \simeq 30$, appropriate for most diffuse ISM situations.

The above result breaks down when λ_T becomes smaller than
λ_e; the heat flux reaches an upper bound which can be estimated
roughly as:

$$F_s \simeq \pi^{-1/2} \ n_e m_e (kT/m_e)^{3/2}$$

$$\simeq 5p(p/\rho)^{1/2}. \qquad\qquad (5.4)$$

The breakdown of the linear regime of equation 5.2 actually
occurs earlier ($\lambda_T \simeq 8\lambda_e$) than naively expected ($\lambda_T \simeq \lambda_e$),
because the bulk of the heat flux is carried by electrons having
velocities around $\sim$ 1.7 times the thermal velocity used in
equation 5.1, and the mean free path rises as u^4.

Even the smallest ($B_{crit} \simeq 10^{-6} \ n \ T^{-3/2}$ gauss) magnetic
field will suppress thermal conduction in directions perpendicular to it. Electron thermal conduction occurs only along the
magnetic field.

Because $\kappa \propto T^{5/2}$, while the thermal energy density $E \propto nT$,
thermal conduction will be most important in high temperature,

low density situations.   An example will be given in section 6.
Thermal conduction   in   neutral gas   is unimportant   in   all   ISM
situations.

6. H$^+$ REGIONS

This and the following sections will put to use the building
blocks of the previous sections in analyzing idealized versions
of situations thought to occur in the diffuse ISM.   Other
examples (supernova remnants and wind bubbles) can be found in
the contribution of H.J.G.L.M. Lamers.

6.1 Static Problem

Assume constant density of neutral gas,   instantaneous turn-
on of ionizing flux to constant value N_L.   We first freeze the
dynamics   and   recall   a   few classical results concerning the
expansion of the H$^+$ region.

$$r(t) = r_s(1 - \exp(-t/t_s))^{1/3} \tag{6.1}$$

$$dr/dt = u_s(r/r_s)^{-2}\exp(-t/t_s) \tag{6.2}$$

$$r_s = (3/4\pi\alpha_B)^{1/3}\ N_L^{1/3}\ n^{-2/3} \tag{6.3}$$

$$u_s = r_s/3t_s \tag{6.4}$$

$$t_s = (n\alpha_B)^{-1} \tag{6.5}$$

where r_s is the Strömgren (equilibrium) radius of the H$^+$ region,
u_s and t_s characteristic velocity and timescale for the formation
of the Strömgrem sphere, and α_B the recombination coefficient to
states above ground (recombinations to ground state regenerate
Ly-c photons).   The Strömgren sphere image is valid if the
transition region neutral/ionized is sharp compared to r_s; this
is true when the parameters τ_s is large:

$$\tau_s = n\sigma_o r_s \tag{6.6}$$

where σ_o is a mean value for the photoionization cross section.
We give numerical formulae and introduce the excitation parameter
U:

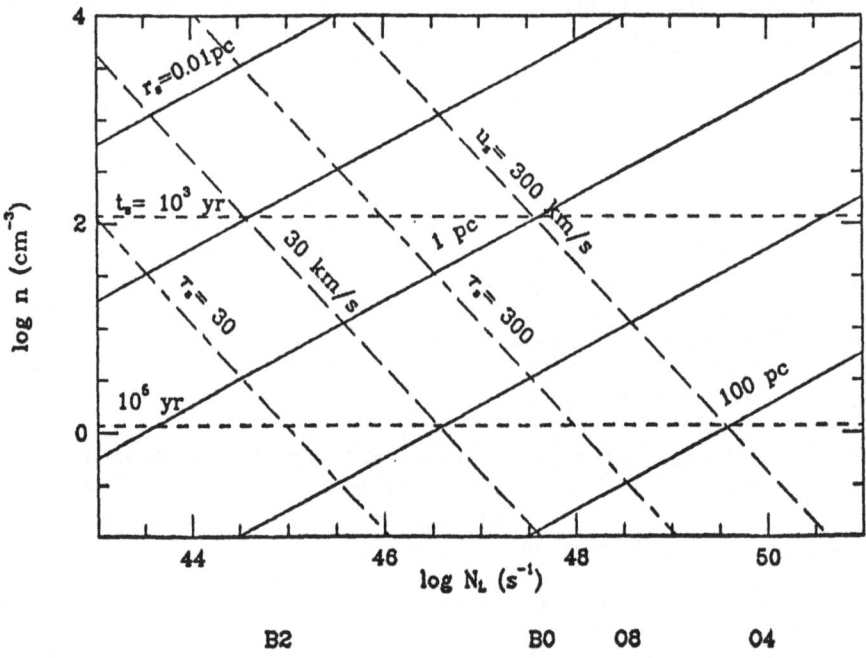

Fig 6.1 Relations between various properties of Stromgren spheres. Radius : full line; formation time : short dash; characteristic velocity of ionization front during formation : long dash; Ly-c optical depth into original neutral gas : long-short dash.

$$r_s/1 \text{ pc} = U(n/1 \text{ cm}^{-3})^{-2/3} \qquad\qquad (6.7)$$

$$\tau_s = 18 \, U(n/1 \text{ cm}^{-3})^{1/3} \qquad\qquad (6.8)$$

$$U = 32 \, (N_L/10^{48} \text{ s}^{-1})^{1/3}. \qquad\qquad (6.9)$$

The value of α_B at $T = 10^4$ K has been used. Figure 6.1 gives an overview of parameter space. Noteworthy points are:

i) $\tau_s \gg 1$ for all cases of interest: H^+ regions have sharp edges.

ii) Generally $u_s \gg c_2$ during the formation phase, justifying our neglect of dynamical effects.

iii) The formation timescale is short, and the formation phase ($t < t_s$) is unlikely to be observed, even more so when the finite turn-on time of the star is taken into account.

6.2 Dynamical Evolution

What if we reinstate dynamics?   Our self consistent assumption of constant volume ionization corresponds to the I-front being exteme-weak R.   If $u_s \gg c_2$, dr/dt will decrease to $2c_2$ only when r is close to r_s.   This occurs as the photon flux at the I-front decreases, and the critcal value $2n_1c_2$ is reached. A shock then appears as explained in section 4, and the I-front becomes D-type.   Note that the near coincidence in time of (a) change of I-front nature, (b) r reaching the Stromgren radius, is not fundamental but arises in commonly encountered situations.   Time dependent dynamical effects are important during that phase and can be treated only numerically [52], [53], [54].   It has been properly pointed out that departures from steady state I-front structure are also important in that phase [50], [51].

Later evolution ($c_1 \ll dr/dt \ll c_2$) can be described analytically.   A strong radiative shock moves into the neutral gas, the shocked gas is compressed in a thin layer by the pressure of the H^+ region, from which it is separated by a wD I-front.   By expressing momentum balance for the shocked layer and recombination balance for the H^+ region, the following asymptotic expansion law [3] can be derived:

$$r = r_s(1 + 7c_2t/4r_s)^{4/7}. \qquad (6.10)$$

Note that the maximum velocity attained by <u>ionized</u> gas during the whole evolution is $c_2 \simeq 10$ km/s.

6.3 Absorption of Ly-c Photons by Dust

For standard dust properties, absorption of Ly-c photons by dust is significant when $\tau_s \geqslant 10^3$ (defined in equation 6.8). This problem has been studied in [55], [56]; a useful analytic approximation for the fraction of Ly-c photons absorbed by the gas is:

$$f = \exp(-n_g\sigma_g r_a) \qquad (6.11)$$

where r_a is the actual radius of the dusty H^+ region.

6.4 Champagne Applied to Blisters. Does It Work?

A series of papers [57 and references therein] have studied the dynamics of H^+ regions in an inhomogeneous ambient medium. The basic configuration is that of an ionizing star embedded in a neutral cloud, which is of finite extent, and pressure confined by a tenuous outer (intercloud) medium.   In the late dynamical evolution of an H^+ region, described above in 6.2, the pressure

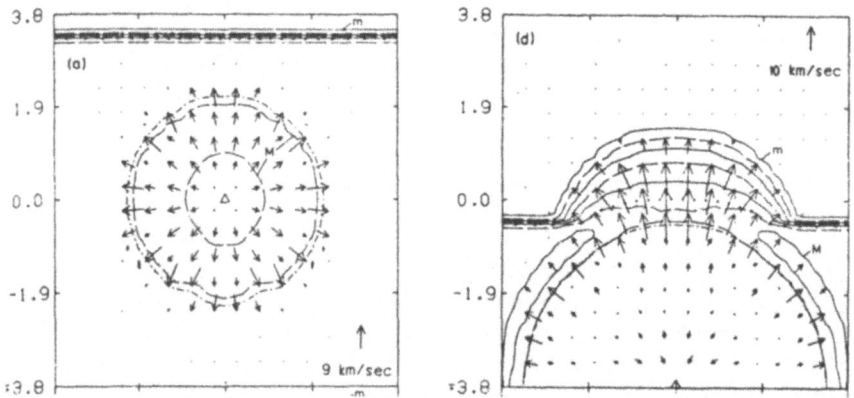

Fig 6.2 Two stages in the evolution of an H^+ region near the edge of a neutral cloud embedded in a tenuous intercloud medium : initial spherical expansion, and "cork ejection". From Bodenheimer, Tenorio-Tagle, and Yorke Ap.J. 233, 85, by permission of the authors.

of the ionized gas is balanced essentially by the ram pressure of the outer medium accreting onto the dense neutral shell. The thermal pressure of the outer medium plays a negligible role. When the expansion of the H^+ region reaches the edge of the cloud, the ram pressure confinement ceases, and the dense shell is ejected into the intercloud medium, followed by the freely expanding ionized gas. Whence the name champagne flow coined by the authors. Figure 6.2 illustrates this.

We now give an approximate analytical treatment of the gas flow in the champagne phase. It is interesting both as an illustration and as a tool to obtain results in cases not covered by published numerical calculations. The geometry is described in figure 6.3. We treat it as a pipe of varying cross-section, and make a crude approximation by writing down equations which would be valid if the scale along the flow direction were large compared with the transverse scale.

$$\rho u \Sigma = A \qquad (6.12)$$

$$u\frac{du}{dx} + \frac{c^2}{\rho}\frac{d\rho}{dx} = 0 \qquad (6.13)$$

$$(m - m^{-1})\frac{dm}{dx} = \frac{dLog\Sigma}{dx} \qquad (6.14)$$

$$\Sigma = \pi(r_0^2 + x^2). \qquad (6.15)$$

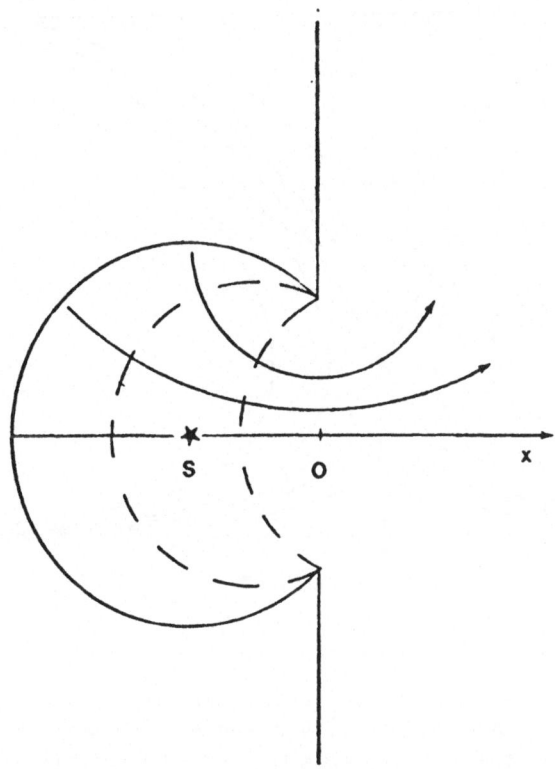

Fig 6.3 Schematic geometry adopted to model an H^+ region freely expanding into the intercloud medium (champagne phase)

Σ is the cross section of the pipe, of which (6.15) gives a reasonably good representation; A is the constant mass flux; $m = u/c$ the isothermal Mach number. Steady state flow has been assumed. Equation (6.14) is the well-known equation for the de Laval nozzle: the flow can make a smooth transition from sub-to supersonic motion only at the throat (minimum Σ). Setting $z = x/r_0$, (6.14) integrates to:

$$(m^2 - 1)/2 - \text{Log } m = \text{Log}(1 + z^2) + B \qquad (6.16)$$

where $B = 0$ for the critical solutions that go smoothly through ($z = 0$, $m = 1$). We assume (see section 4.4) that only weak or critical I-fronts can occur. We also assume that the champagne bottle pops open after the initial formation of the Strömgren sphere, so that the star is unable to drive an R-type I-front into the cloud. A few sample solutions are drawn on figure 6.4.

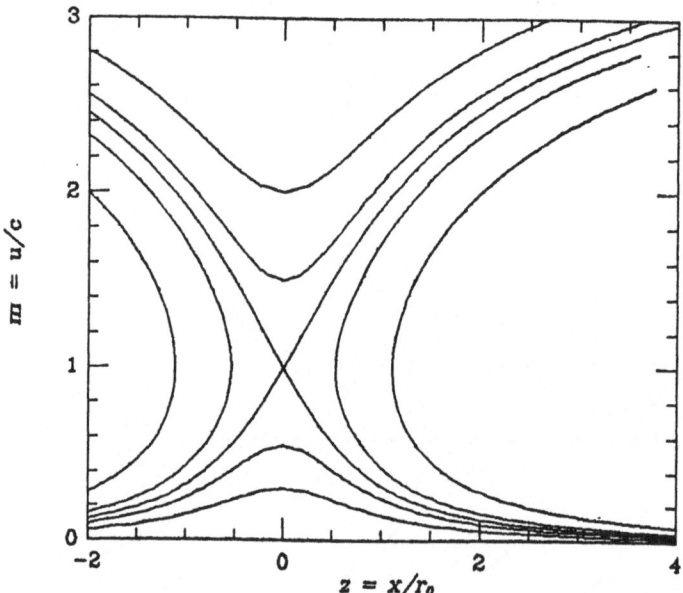

Fig 6.4 Representative solutions of eq. 6.14, showing
isothermal Mach number versus reduced distance z. Solutions
start from the left at an abscissa corresponding to the wall
of the H^+ region, according to the geometrical configuration
under consideration.

Neglecting the expansion velocity of the cavity wall (consistent
with the assumption of steady state flow), the Mach number at the
cavity wall ($z_0 = x_0/r$, negative) must be the exhaust Mach number
of a wD or cD I-front, thus subsonic.

If we start above the subsonic critical branch, 'the solution
does not reach the throat. If we start below the subsonic
critical branch, the solution remains subsonic and has asymptoti-
cally constant pressure, which is in general incompatible with
the intercloud pressure. We therefore must pick the critical
solution, which becomes supersonic beyond the throat. For
definiteness, we now pick a particular geometry (for which figure
6.3 is drawn), with cavity radius $r_c = 2^{1/2}r_0$, and the cavity
wall at $z_c = -(1 + 2^{1/2})$; then, $m_c \simeq 0.21$. The velocity law is
determined, and, using equation 6.12, the density is determined
except for a scale factor that must be found. We write down the
budget for Ly-c photons within the solid angle subtended by the
cavity from the star, using the fact that, for $x_c < x < x_*$
($x_* = -1$, abscissa of star), the product $m\Sigma$ is constant within
5%, and therefore so is the density

$$N_L = \pi r_c^2 n_c cm_c + (4/3)\Pi r_c^3 n_c^2 \alpha_B. \tag{6.17}$$

The solid angle cancels out, and the only dependance on the geometry is through $m_c = m(x_c)$. We now guess that, for large enough N_L, the second term on the right is dominant, which happens when:

$$n_c > 3cm_c/4\alpha_B r_c \tag{6.18}$$

$$N_L > 3_c^2 m_c^2 \Pi r_c/2\alpha_B. \tag{6.19}$$

The mass flow rate is then:

$$A = \Omega \ \mu_e cm_c \ (3N_L r_c/4\alpha_B)^{1/2} \tag{6.20}$$

where Ω is the solid angle subtended by the cavity; $\Omega = \Pi(2 + 2^{1/2})$ for the geometry that we chose; and μ_e is the mass per electron in the ionized gas. We now make a comparison with the numerical calculation of [58], case 2, model f, which has a similar geometry. Adopting the values given in the text by the authors or read off their figure 3f: $N_L = 7.6 \times 10^{48}$ s^{-1}, $\alpha_B = 3.5 \times 10^{-13}$, $\mu_e = m_H = 1.66 \times 10^{-24}$ g, $r_e \approx 4.2 \times 10^{18}$ cm, we compute a mass flow rate:

$$A = 1.6 \times 10^{22} \ g \ s^{-1}$$

$$= 2.5 \times 10^{-4} \ M_0 \ yr^{-1}.$$

The mass flow rate of 3×10^{-3} M_0 yr^{-1} quoted by the authors in the conclusion does not apply to their case 2; an approximate integration of the mass flow across the isodensity contour 10^{-22} g cm^{-3} of their figure 3f gives: $A = 1.1 \times 10^{-4}$ M_0 yr^{-1}. The factor of ~ 2 difference with our analytical result can be explained i) by our neglect of dust absorption, ii) by the fact that the steady state flow has not yet developed in the numerical calculation: less than 5×10^4 yrs have elapsed since the champagne flow began, the sound travel time across the cavity is 2.8×10^5 yr, and the 07 star still has several 10^6 yr to live with slowly decreasing Ly-c flux. The assumption of steady-state flow will become better as evolution proceeds and the expansion of the cavity slows down.

The total mass ionized by an 0 star in the champagne process is much larger than what it could have ionized (for the same time and cloud density) in a bounded H$^+$ region. Supersonic motions are generated in the champagne flow; however, the velocity rises slowly beyond the throat: $m \approx 2(\text{Log } r/r_0)^{1/2}$, while the emissivity decreases sharply $n^2 \approx n_0^2 (r/r_0)^{-4}$; supersonic motions may be hard to detect.

The champagne model is thought to correspond to the objects known as blisters: compact, asymetric, and often obscured H^+ regions found associated with dense clouds. Comprehensive observations of Sharpless 206 are described in [59]; [60] gives a review of observations. Some support to the model is given by the fact that the radial velocity difference $\Delta v = v_{HII} - v_{CO}$ between ionized and neutral gas tends to be more negative for visible blisters (on the side of the cloud facing the observer) than for obscured nebulae. A detailed modeling of the infrared and radio continuum emission of 30 Doradus [61] gives good agreement with the observations.

In a related class of objects, the ionizing star is outside the cloud, giving rise to bright rims, elephant trunks, etc. [62]. Dynamical models have been presented, e.g. [63].

7. CLOUDS UNDER HARDSHIP

We describe in this section several kinds of aggression that clouds can undergo.
7.1 Shocks

What happens when a shock wave (such as the blast of a supernova remnant, or a spiral wave shock) propagating in a tenuous intercloud medium overtakes a denser cloud? If the cloud was initially at rest with respect to the intercloud medium, and the shock is strong, the post shock flow around the cloud is mildly supersonic: $M_{ad} = 3/5^{1/2} \simeq 1.34$ (note the distinction with the result given in section 3.1). A reverse bow shock therefore appears in the intercloud gas. Besides, the impulsive pressure increase drives a converging shock into the cloud; because of the relative motion of the cloud and intercloud, the pressure at the surface of the cloud is not constant, and the cloud shock is strongest on the upwind side. The resulting configuration is schematized in figure 7.1.

Values behind the primary shock are:

$$v_1 = (3/4)v_b \qquad\qquad\qquad\qquad\qquad (7.1)$$

$$p_1 = (3/4)\rho_0 v_b^2. \qquad\qquad\qquad\qquad (7.2)$$

Further compression occurs as the gas goes through the bow shock and comes to rest against the cloud surface;

$$p_2 = 2p_1 \qquad\qquad\qquad\qquad\qquad\qquad (7.3)$$

$$p_3 \approx 3p_1. \qquad\qquad\qquad\qquad\qquad\qquad (7.4)$$

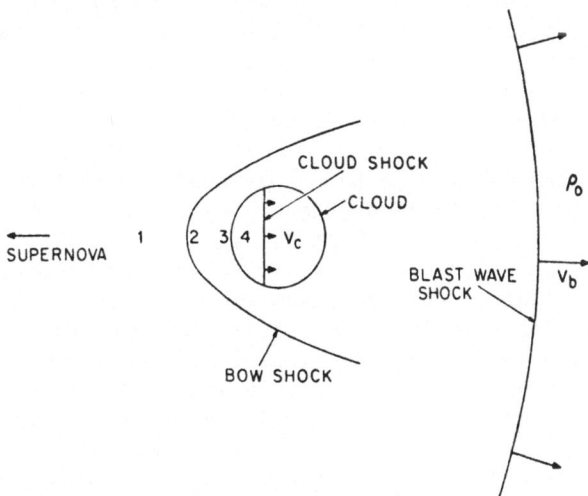

Fig 7.1 Schematic representation of the shock waves that occur when a blast wave produced by a supernova overtakes a cloud. The gas in region 1 is the hot gas behind the blast wave, region 2 is just behind the bow shock, point 3 is the stagnation point at the surface of the cloud, and region 4 contains the gas which has passed through the cloud shock. the cloud shock moves at velocity v_c; the blast wave shock moves at velocity v_b into the ambient medium of density ρ_b. Reproduced from McKee and Cowie 1975 Ap.J. 195, 715, by permission of the authors.

Expressing momentum conservation across the cloud shock:

$$P_3 \approx \rho_c v_c^2 \qquad (7.5)$$

$$v_c \approx (3\rho_0 / 2\rho_c)^{1/2} \, v_b. \qquad (7.6)$$

For young SNR's, radiation is dynamically unimportant in the main and bow shocks. The cloud shock has both lower velocity and higher density, and therefore is radiative if ρ_c is large enough (7.4 then becomes exact). More details can be found in [64], [65]. Besides being shocked, the cloud is accelerated outwards by the drag of the flow [66], and both Rayleigh-Taylor and Kelvin-Helmholtz instabilities can develop at the surface of the cloud [67]. This model nicely resolves some discrepancies in observed properties of SNS's:

-two classes of optically emitting objects in Cas A: the "fast moving knots" with velocities ~ 5000 km s^{-1}, and the "quasi stationary floculi" having v ~ 500 km s^{-1}.

-difference between expansion velocity deduced from X-ray observations (e.g. $v_b \simeq (16 \, kT/3\mu)^{1/2}$) and radial velocities from optical observations [68], [69], [70]. Both sets of observations can be reconciled consistently if the low radial velocity optical emission is attributed to radiative shocks moving into the clouds.

A related model has been proposed to explain the optical emission from Herbig-Haro objects [71], [72]: a high density, moderate velocity ($n_w \simeq 300 \, cm^{-3}$, $v_w \simeq 100 \, km \, s^{-1}$) stellar wind impinges upon a dense cloudlet. Because of the higher density and lower velocity, the bow shock (as well as the cloud shock) is radiative. Emission from both shocks is observed (optical atomic lines for the bow shock, H_2 vibrational transitions in the near IR for the cloud shock), and independent determinations of the post-shock pressures (see eq. 7.2 and 7.4) are in satisfactory agreement: $n_c v_c^2 \simeq n_w v_w^2$, with $n_c \simeq 10^4 \, cm^{-3}$ and $v_c \simeq 15 \, km \, s^{-1}$. Note that pressure balance and $n_c \gg n_w$ imply $n_w v_w^3 > n_c v_c^3$; i.e. the bow shock is the most luminous.

7.2 Ionizing Flux

We now consider the response of a neutral cloud to an external flux of Ly-c radiation. In [74], the case of an anisotropic ionizing flux is studied, and it is proposed that the precursor shocks of D-type fronts might be the main acceleration for interstellar clouds, by the so-called rocket effect. In [75], an isotropic flux is considered; the dynamics and ionization budget are studied in detail. The notion of a recombination jacket consuming most of the incoming Ly-c flux, is introduced (we used a similar approximation in section 6.4). Note, however, that the results of [75] cannot be applied as they stand if, as argued in section 6, the exhaust Mach number of the I-front is 1; in that special case, some approximations must be reconsidered. The rate of mass loss for a cloud exposed to the flux of a hot star can be expressed as:

$$dM/dt = 3 \times 10^{-4} \, r_c^{3/2} \, N_{48}^{1/2} \, d^{-1} \, (1-(d/d_s)^3)^{1/2} \qquad (7.6)$$

where dM/dt is in $M_0 \, yr^{-1}$; r_c is the cloud radius in parsecs, d and d_s the distance to the star and the radius of its Strömgren sphere, and N_{48} the Ly-c output of the star in $10^{48} \, s^{-1}$. Note that the mass flux varies as the square root of the photon flux, because recombinations dominate the photon budget, as long as:

$$25 \, S_{48}^{1/2} \, r_c^{1/2} \, d^{-1} \gg 1. \qquad (7.7)$$

When an O star is located within an inhomogeneous interstellar medium, the development of an extended H^+ region in the tenuous intercloud is modified and hampered [76]. The emission of ionized gas by the denser clouds "chokes" the low density Strömgren sphere. This limits the possibilities for maintaining a diffuse ionized intercloud component by isolated O stars.

7.3 Evaporation

If a cloud is immersed in a warmer intercloud medium, a heat flux flows along the temperature gradient. If radiative losses can be neglected, this heat flux increases the temperature of the colder gas that expands and eventually merges into the inter- cloud. If one assumes that classical conduction prevails (see section 5) and that gas motions are subsonic, the temperature profile and mass flux of the steady-state flow can be easily derived. The conductive flux just balances the enthalpy flux:

$$4\pi r^2 \kappa(T) dT/dr = (5/2)(kT/\mu) dM/dt \qquad (7.8)$$

so that:

$$dM/dt = (16\pi/25k)\kappa(T_0)\mu r_c \qquad (7.9)$$

$$= 0.4 \ M_0 \ yr^{-1} \ T_{0,6}^{5/2} \ r_{c,p} \qquad (7.10)$$

where $T_{0,6}$ is the unperturbed ambient density in $10^6 K$ and $r_{c,p}$ the cloud radius in pc.

Both assumptions of classical condutions and subsonic flow break down when

$$\sigma > \phi_s$$

where:

$$\sigma = 4.2 \times 10^{-3} \ T_{0,6}^2/n_0 r_c \qquad (7.11)$$

and ϕ_s is a number of order unity measuring the uncertainty of the saturated flux (eq. 5.4). Evaporation rates in the classical and saturated cases are derived in [77]. For $\sigma > 100$, hot electrons can reach directly to the surface of the cloud in one mean free path; the corresponding regime is studied in [79].

Radiative losses become dominant for low values of σ; for $\sigma < 0.03$, the outflow is replaced by condensation. The role of radiative losses is treated in detail in [78].

8. OBSERVATIONAL ANCHORS

We present here observational results that give valuable leads or key constraints when one tries to build a consistent picture of the diffuse ISM. We shall devote more space to the most "modern" component of the ISM: the hot, or coronal, phase.

8.1 Neutral Gas

Neutral gas contains most of the mass of the ISM. We list below the various parameters that define the state of the neutral gas, from the best to the least precise determinations.

Mean density. $\langle n \rangle \simeq 1$ cm^{-3} (solar neighborhood within ~ 1 kpc), splitting roughly as follows: $2\langle n(H_2)\rangle > \simeq 0.5$ cm^{-3}, from UV absorption lines and dust extinction.
$\langle n(HI)\rangle \simeq 0.5$ cm^{-3}, from 21 cm emission line, out of which $\sim$ 2/3 is "cold" and 1/3 is "warm" [80], [81].

Temperatures range 30-500 K for the "cold" gas, with 80 K being a common value, determined from 21 cm emission and absorption and from populations in the H_2 rotational levels. For the "warm" gas, one has in most cases only a lower limit to the temperature $T > 1000$ K, say, but probably $T \sim 10^4$ K because of the sharp rise of the cooling curve at that temperature, irrespective of detailed models. See reviews in [82], [83].

Pressures can be derived from the H_2 rotational or the CI fine structure level populations [84], [85], and M. Jura's contribution. A typical range of values is: 10^3 cm^3 K $< P/k <$ 10^4 cm^3 K.

Spatial structure information is seldom complete. It can be: frequency distribution of column densities N_H on a sample of lines of sight [86], distribution of cloud diameters or column densities over a sample of clouds [87], [88], distribution of 21 cm optical depth [82] or column density [89] over a sample of lines of sight. One or more assumptions concerning the geometry of the "clouds," a constant density, temperature, or pressure, have to be made to relate the various types of observations among themselves and to ideal descriptions of the "cloud" population. Besides, various techniques have different sensitivities. Well established : cloud frequency on a line of sight $\nu \simeq 5$-10 kpc^{-1}; cloud filling factor $f \simeq 0.015$; more uncertain, the cloud spectrum, often schematized as $d\#/dm \propto m^{-\beta}$, where $\beta = 1.5$-2. The warm phase, objectively described as having low 21 cm optical depth, was a few years ago thought to be pervasive (i.e. to have a space filling factor close to unity). Recent evidence (see further) has given credit to the idea that the coronal phase might be the most pervasive. Observations of the warm neutral gas are still with us, and this phase, while it may or may not be pervasive [92], is widespread and can be confined to a small filling factor only by high pressure.

Kinematics. Velocity dispersion $\langle v_r^2 \rangle^{1/2} \simeq 6$ km s^{-1} [82], [90], but some evidence for a high-dispersion (35 km s^{-1}) population of warm (~ 300 K) clouds [91] which might carry the injection of kinetic energy into the general population.

Magnetic field measurements, using the Zeeman splitting of the 21 cm line (in situ) or Faraday rotation (line of sight n_e-weighted average), are very delicate. Numbers from the litterature may be subject to a selection effect favoring detectable fields. See [92] for sensitive measurements and upper limits, showing that magnetic energy is at most twice the thermal energy in diffuse clouds, and certainly less than bulk kinetic energy. These results give an excuse for ignoring the complications of the magnetic field in diffuse ISM dynamics (see, however, sections 3.3 and 3.4).

Energy expenditures have until recently been estimated indirectly, from the cooling function (section 2.3). The main problem, being to make good for these losses by some source of energy, could be attacked from both ends; i.e. one could reduce the assumed losses by decreasing carbon abundance in the gas phase, or the gas pressure. A nearly direct measurement of the cooling rate through the 156 μ line of C$^+$ gives rather high values of the cooling rate per H atom [96]:

$$\ell \simeq 0.7\text{-}3 \times 10^{-25} \text{ erg s}^{-1}$$

which can probably be accomodated only by grain photoelectron heating [97], [98], [99], with low threshold and fairly high yield [100], [101].

8.2 Diffuse Ionized Gas

Measurements of pulsar dispersion measure:

$$DM = \int_0^* n_e \, d\ell \qquad\qquad (8.1)$$

coupled with estimates of pulsar distance L from 21 cm absorption [92], [93], [94], lead to estimates of $\langle n_e \rangle = DM/L$ which are generally close to: $\langle n_e \rangle \simeq 0.025$ cm^{-3}. On the other hand, observations of diffuse Hα [95] give access to:

$$EM = \int_{0}^{\infty} n_e^2 \, d\ell . \qquad\qquad\qquad (8.2)$$

The upper bound of this line of sight integral is different from that giving DM, and can be determined either by the thickness of the ionized gas region ($\sim$ 500 pc, [90]) or by galactic extinction. Combining both types of observations leads to limits/estimates for the local electron density and filling factor:

$$n_e \lesssim 0.3 , \qquad\qquad f \gtrsim 0.1.$$

Warning: our notation is different from that of [90], and read further down a discussion of the definition(s) of "f".

8.3 Coronal Gas: OVI Data

One of the few theoretical predictions in astronomy was that of "coronal" gas at temperatures of the order of 10^6 K [102]. Confirmation came with the observation by the Copernicus satellite of the 1032–1037 Å resonance doublet of OVI in absorption in front of hot stars. The equilibrium ionic abundance of OVI peaks at $\log T \simeq 5.5$, so the observations show that somewhere on the line of sight, there is gas at roughly that temperature. The merged set of observations is published in [103], and an extensive discussion in [104], of which we shall outline the main arguments.

The discussion of [104] is based on the results of 72 lines of sight to OB stars, mostly within 2 kpc. The absorption lines are shallow, their width is resolved by the instrument, but they otherwise show no structure. The information for each line of sight is therefore embodied in a set of three numbers: N_i, column density of OVI; v_i, velocity centroid; σ_i, rms velocity dispersion. The N_i are correlated with the star's distance, but with a rather large scatter. The author addresses the question of the smoothness of the emitting medium by assuming it to be made up of individual "chunks," having individual column density N_0, internal and external velocity dispersion σ_0 and σ respectively. Two independent approaches are taken. First the statistics of the velocity dispersion is used. The total sample being large, one expects the realization of the sample velocity variance to be close to its expectation value:

$$\langle v_i^2 \rangle + \langle \sigma_i^2 \rangle \simeq \sigma_0^2 + \sigma^2 .$$

But the statistics of variances σ_i^2 on individual lines of sight depends on the mean number of chunks

$$\langle \# \rangle = \langle N_i \rangle / N_0$$

on a line of sight. If $\langle \# \rangle$ is large, each line of sight is a fairly good poll of the whole sample, and one expects: $\sigma_i^2 \simeq \sigma_0^2 + \sigma^2$. If $\langle \# \rangle$ is of order unity, a wider distribution $\sigma_0^2 < \sigma_i^2 < \sigma_0^2 + \sigma^2$ is expected. Such is the flavor of the method used to derive: $N_0 \simeq 1.3 \times 10^{13}$ cm^{-2}. A similar, but independent method based on the statistics of the N_i gives a closely similar result: $N_0 \simeq 1.05 \times 10^{13}$ cm^{-2}. The smoothed out density of OVI is $\langle n(OVI) \rangle \simeq 2.8 \times 10^{-8}$ cm^{-3}, and the line of sight frequency of the individual components used in the statistical analyses is: $\nu \simeq 6$ kpc^{-1}, suggestively close to that of cold clouds. Careful tests show that the observed OVI is not connected in any way to the background stars used as UV point sources.

We may now ask: do we have room for the gas where the OVI lines are formed? A simple parametric representation assumes:

$$d\langle n_e \rangle = g(T) \; dLogT \qquad\qquad (8.3)$$

where the l.h.s. measures how much gas, in a unit volume of interstellar space, is present in a given interval of Log T. Note that the n_e in 8.3 has nothing to do with the actual local density of the gas. In several papers discussing this problem, $g(T) = CT^\eta$ for $T_1 < T < T_2$. The assumed distribution must reproduce the observed mean $n(OVI)$:

$$\langle n(OVI) \rangle = \int \frac{n(OVI)}{n(O)} \cdot \frac{n(O)}{n_e} \, d\langle n_e \rangle \; . \qquad\qquad (8.4)$$

If the rates in the integral are a function of temperature only, eq. 8.4 directly determines the normalization C. We now know how much stuff we need to account for the OVI; how much space does it take up? We must specify the local density as a function of temperature, say:

$$n_e = h(T). \qquad\qquad (8.5)$$

The volume filling factor f is given by:

$$f = \int n_e^{-1} \, d\langle n_e \rangle = \int [h(T)]^{-1} \, g(T) \; dLogT. \qquad\qquad (8.6)$$

For problems involving n_e^2, one may wish to define the clumping factor:

$$c = \frac{\langle n_e^2 \rangle}{(\langle n_e \rangle)^2} = \int n_e d\langle n_e \rangle / [\int d\langle n_e \rangle]^2$$

$$= \int h(T)g(T)dLogT / [\int g(T)dLogT]^2 \,. \tag{8.7}$$

Note that, contrary to what is stated in [104], $f \neq c^{-1}$, whenever a range of temperature and densities in involved. If we assume pressure equilibrium,

$$P/k \simeq 2n_e T \tag{8.8}$$

$$f = 2(P/k)^{-1} \ C\int T^{\eta+1} \ dLogT. \tag{8.9}$$

The normalization C being determined by eq. 8.4, we obtain a constraint on the product $f \cdot (P/k)$. Because $f < 1$, we have an absolute lower bound on (P/k); the actual constraint is more severe because other phases have legitimate claims to fill a sizeable portion of interstellar space: diffuse ionized, warm neutral, etc. For instance, if $\eta = 0.5$ for $4.7 < Log\ T < 6.3$, then:

$$f \cdot (P/k) = 1.8 \times 10^3 \ cm^3 \ K \tag{8.10}$$

which can be accomodated with reasonable values of f and (P/k).

8.4 Coronal Gas: Soft X-ray Data

     Various groups have carried out photometric measurements of the background flux in various energy bands of the EUV-SXR spectrum from 0.1 to 1 keV. The emission is of galactic origin, patchy, and shows no clear correlation of either sign with HI column densities [105], [106], [107]. The thermal nature of the emission mechanism is confirmed by observations of line emission [108]. The observations are interpreted by various authors within the framework of simple assumptions for the spatial and temperature distribution of the emitting gas.

     Single-temperature models. The spatial distribution can be bracketed between two extreme types; the absorbed slab, for which the received intensity:

$$I(E,T) = \frac{1}{4\pi} f \ n_e^2 \ \Lambda(E,T) \ R \ exp[-N_H \sigma(E)] \tag{8.11}$$

where the emitting gas fills a fraction f of a line of sight of length R behind a column density N_H of neutral hydrogen; and the interspersed, or emulsion, model:

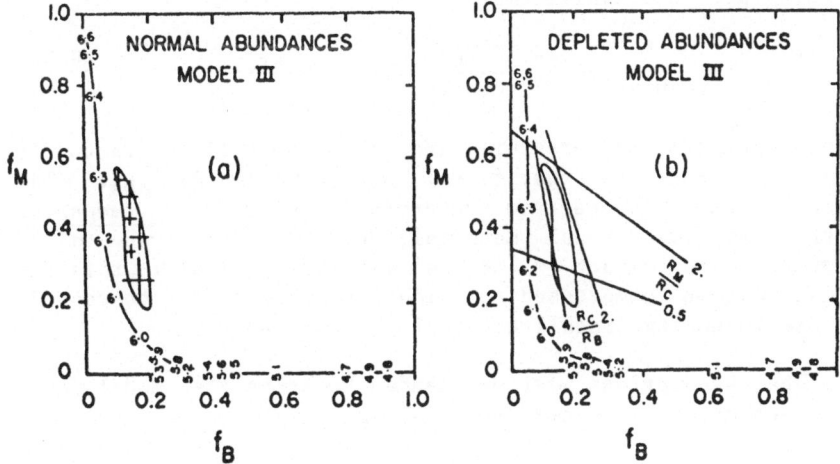

Fig 8.1 Color-color diagram for diffuse soft X-ray intensities. Three energy bands are oserved, called B, C, and M in order of increasing energy. The band fractions f_M and f_B are the ratio of the count rates in the respective bands to the total count rate B+C+M. Reproduced from Burnstein, Borken, Kraushaar, and Sanders Ap.J. 213, 405 [105], by permission of the authors.

$$I(E,T) = \frac{f \, n_e^2 \, \Lambda(E,T)}{4 \, \pi \, n_H \, \sigma(E)} \qquad (8.12)$$

where the emitting gas is interspersed on a small scale with neutral gas having a mean density n_H. We now outline the discussion used in [105]. For each model, let the temperature vary. Intensity ratios for two pairs of energy bands define a trajectory in the color-color diagram; the analogue of the U-B, B-V blackbody curve. Observational points do not fall on the curve; therefore, a superposition of emission from at least two temperatures is necessary. See figure 8.1. Assume we choose the interspersed model. The color-color diagram constrains the shape of the temperature distribution $d\langle n_e\rangle/d\text{LogT}$ within a scale factor. But this scale factor is fixed by the mean density of OVI (eq. 8.4). We now require that we reproduce not only the spectrum, but the absolute intensity of the soft X-rays at any energy:

$$I(E) = \frac{1}{4\pi n_H \sigma(E)} \int n_e \frac{d\langle n_e\rangle}{d\text{LogT}} \Lambda(E,T) \, d\text{LogT}. \qquad (8.13)$$

Assuming pressure equilibrium:

$$I(E) = \frac{P}{k8\pi n_H \sigma(E)} \int T^{-1} \frac{d\langle n_e \rangle}{dLogT} \Lambda(E,T) \, dLogT. \qquad (8.14)$$

This determines the pressure, and thus the filling factor (eq. 8.6). It turns out that $f \ll 1$ and $(P/k) \gg (P/k)_{obs}$. In a nutshell, the SXR spectrum determines the relative temperature distribution of coronal material; then the observed $\langle n(OVI) \rangle$ determines the amount of coronal material; finally this fixed amount must be compressed to high energy pressure to reproduce the SXR intensity (recall emissivity/atom $\propto n_e$).

What assumptions must be amended to bring the model results in agreement with the observations?

Pressure equilibrium need not be achieved; indeed, in any model where supernovae are the energy sources, the hottest parts of the ISM (young SNR's) also have higher pressure, and maybe higher density as well.

Ionization structure may depart from equilibrium; either in shock-heated gas or in evaporation fronts around cold clouds. Note that some analyses of OVI-SXR data make inconsistent use of non equilibrium cooling calculations [17] when applying them to evaporating clouds, where the ionization departure from equilibrium should be in the opposite direction.

How typical is our vantage point at the Sun? There is some evidence [105], [109] that the Sun might be inside a bubble of radius R $<$ 100 pc, temperature T $\simeq 10^6$ K, and pressure P/k $\simeq 10^4$ cm^{-3} K, above average. Also, the mean HI density is lower $n_H \simeq 0.1$ cm^{-3} within the immediate neighborhood ($\sim$ 100 pc). Likewise, soft X-ray observations, especially at low energies, sample a smaller volume than OVI absorption.

Stellar contributions to the soft X-ray background, although significant, are not dominant, and cannot be used to explain away the problem met above [110].

We have outlined some of the problems met, when trying to find a set of physical conditions for the interstellar gas, that agree with a set of observations. Global models attempt to produce these physical conditions consistently from a minimal set of assumptions, e.g. total mean gas density $\langle \rho_g \rangle$ and supernova rate S. Two such models are presented in [111] and [112,113]. See also reviews in [114] and [115]; a modeling of the local hot bubble in [116]; and investigations of the time-dependant, large-scale evolution of the ISM in [117] and [118].

CONCLUSION

We have essentially tried to provide the reader with a description of usable tools. We have highlighted some "economical approximation" ideas, like the regimes of radiative cooling, the "extreme" varieties of ionization fronts, the "isothermal" shocks. But at the same time, we have pointed out the limits of the domain of validity of each of these convenient approximations. In the last section, we have hardly scratched the problem of modeling the ISM at large.

ACKNOWLEDGEMENTS

S. White and C. McKee gave useful advice on preliminary versions of this text. This work supported by grant AST 79 23243.

REFERENCES

[1] Landau, L.D. and Lifshitz, E.M. Fluid mechanics, Pergamon (Oxford) 1959
[2] Zel'dovich, Ya.B. and Raizer, Yu.P. Physics of shock waves and high-temperature hydrodynamic phenomena, Academic Press (New York) 1966
[3] Spitzer, L. Physical processes in the interstellar medium, John Wiley (New York) 1978
[4] Dyson, J.E. and Williams, D.A. Physics of the interstellar medium, Manchester University Press (Manchester) 1980
[5] Patterson, G.N. Molecular flow of gases, John Wiley and Sons (New York) 1956
[6] Wu, T.-Y. Kinetic equations of gases and plasmas, Addison-Wesley (Reading, Massachusetts) 1966
[7] Tritton, D.J. Physical fluid dynamics, Van Nostrand Reinhold (New York) 1977, Chapter 3
[8] Prandtl, L. Essentials of fluid dynamics, Blackie (London) 1967
[9] Oswatitsch, K. Gas dynamics, Academic Press (New York) 1956.
[10] Baines, M.J., Williams, I.P. and Asebiomo, A.S. 1964, M.N.R.A.S. 130, 68
[11] Henderson, C.B. 1976, A.I.A.A. Journal 14, 707
[12] Gail, H.P. and Sedlmayr, E. 1979, A&A 76, 158
[13] Osterbrock, D.E. Astrophysics of gaseous nebulae, Freeman (San Francisco) 1974

[14] Werner, M.W.   1970, Ap.Lett. 6, 81
[15] Walmsley, M.   1973, A&A 25, 129
[16] Dalgarno, A. and McCray, R.A.   1972, Ann. Rev. Astr. Ap.,
     10, 375
[17] Shapiro, P.R. and Moore, R.T. 1976 Ap.J. 207,460
[18] Raymond, J.C., Cox, D.P., and Smith, B.W.   1976, Ap.J.
     204, 290
[19] Raymond, J.C. and Smith, B.W.   1977, Ap.J. Suppl. 35, 419
[20] Cox, D.P. and Daltabuit, E.   1971, Ap.J. 167, 113
[21] Cox, D.P. and Tucker, W.H.   1969, Ap.J. 157, 1157
[22] Gould, R.J. and Thakur, R.K.   1970, Ann. Phys. 61, 351
[23] Lotz, W.   1967, Ap.J. Suppl. 14, 207
[24] Aldrovani, S.M.V. and Péquignot, D.   1973, A&A 25, 141
[25] Aldrovani, S.M.V. and Péquignot, D.   1976, A&A 67, 321
[26] Shull, M.J. and Steenberg, M.V.   1982, Ap.J. Suppl. 48, 95
[27] Ostriker, J.P. and Silk, J.   1973, Ap.J. Lett. 84, L133
[28] Burke, J.R. and Silk, J.   1974, Ap.J. 190, 1
[29] Dwek, E. and Werner, M.W.   1981, Ap.J. 248, 138
[30] Silk, J. and Burke, J.R.   1974, Ap.J. 190, 11
[31] Dwek, E.   1981, Ap.J. 246, 430
[32] McKee, C.F. and Hollenbach, D.J.   1980, Ann. Rev. Astr. Ap.
     18, 219
[32a Draine, B.T. 1980, Ap.J. 241, 1021
[33] Cox, D.P. 1972, Ap.J. 178, 143
[34] Dopita, M.   1977, Ap.J. Suppl. 33, 437
[35] Raymond, J.C.   1979, Ap.J. Suppl. 39, 1
[36] Kafatos, M.   1973, Ap.J. 182, 433
[37] Shull, J.M. and McKee, C.F.   1979, Ap.J. 227, 131
[38] Kahn, F.D.   1954, Bull. Astr. Inst. Neth. 12, 187
[39] Axford, W.I.   1961, Phil. Trans. Roy. Soc. London A253, 301
[40] Giuliani, J.L.   1980, Thesis, Yale University
[41] Mason, D.J.   1980, A&A 92, 117
[42] Axford, W.I.   1964, Ap.J. 140, 112
[43] Newman, R.C. and Axford, W.I.   1968, Ap.J. 153, 595
[44] Giuliani, J.L.   1979, Ap.J. 233, 280
[45] Marsh, M.C.   1970, M.N.R.A.S. 147, 95
[46] Giuliani, J.L.   1981, Ap.J. 245, 903
[47] Tenorio-Tagle, G.   1976, A&A 53, 411
[48] Spitzer, L.   Physics of fully ionized gases, Interscience
     (London) 1962
[49] Hill, J.G. and Marsh, M.C.   1972, M.N.R.A.S. 156, 189
[50] Vandervoort, P.O.   1965, Ap.J. 142, 507
[51] Vandervoort, P.O.   1965, Ap.J. 142, 521
[52] Mathews, W.G.   1965, Ap.J. 142, 1120
[53] Lasker, B.M.   1966, Ap.J. 143, 700
[54] Vandervoort, P.O.   1966, Ap.J. 146, 104
[55] Petrosian, V., Silk, J., and Field, G.B.   1972, Ap.J. Lett.
     177, L69
[56] Mezger, P.G., Smith, L.F., and Churchwell, E.   1974, A&A
     32, 269

[57] Yorke, H.W., Bodenheimer, P., and Tenorio-Tagle, G. 1982, A&A 108, 25

[58] Bodenheimer, P., Tenorio-Tagle, G., and Yorke, H.W. 1979, Ap.J. 233, 85

[59] Deharveng, L., Israel, F.P., and Maucherat, M. 1976, A&A 48, 63

[60] Israel, F.P. 1978, A&A 70, 769

[61] Icke, V., Gatleg, T., and Israel, F.P. 1980, Ap.J. 236, 808

[62] Pottasch, S. 1956, Bull. Astr. Inst. Neth. 13, 77

[63] Sanford, M.T., Whitaker, R.W., and Klein, R.I. 1982, Ap.J. 260, 183

[64] McKee. C.F. and Cowie, L.L. 1975, Ap.J. 195, 715

[65] Sgro, A.G. 1975, Ap.J. 197, 621

[66] McKee, C.F., Cowie, L.L., and Ostriker, J.P. 1978, Ap.J. Lett. 219, L23

[67] Woodward, P.R. 1976, Ap.J. 207, 484

[68] Woodgate, B.E., Kirshner, R.P., and Balou, R.J. 1977, Ap.J. Lett. 218, L129

[69] Galas, C.M.F. and Venkatesan, D. 1981, Ap.J. 250, 216

[70] Rosado, M. 1981, Ap.J. 250, 222

[71] Schwartz, R.D. and Dopita, M.A. 1980, Ap.J. 236, 543

[72] Schwartz, R.D. 1981, Ap.J. 243, 197

[73] Elias, J.H. 1980, Ap.J. 241, 728

[74] Oort, J. and Spitzer, L. 1955, Ap.J. 121, 6

[75] Kahn, F.D. 1969, Physica 41, 172

[76] Elmegreen, B.G. 1976, Ap.J. 205, 405

[77] Cowie, L.L. and McKee, C.F. 1977, Ap.J. 211, 135

[78] McKee, C.F. and Cowie, L.L. 1977, Ap.J. 215, 213

[79] Balbus, S.A. and McKee, C.F. 1982, Ap.J. 252, 529

[80] Guibert, J., Lequeux, J., and Viallefond, F. 1978, A&A 68, 1

[81] Gordon, M.A. and Burton, W.B. 1976, Ap.J. 208, 346

[82] Crovisier, J. 1981, A&A 94, 162

[83] Dickey, J.M., Salpeter, E.E., and Terzian, Y. 1979, Ap.J. 228, 465

[84] Jenkins, E.B. and Shaya, E.J. 1979, Ap.J. 231, 55

[85] Jura, M. 1975, Ap.J. 197, 575

[86] Hobbs, L.M. 1974, Ap.J. 191, 395

[87] Knude, J. 1981, A&A 97, 380

[88] Knude, J. 1981, A&A 98, 74

[89] Dickey, J.M. and Crovisier, J. 1982, preprint

[90] Falgarone, E. and Lequeux, J. 1973, A&A 25, 253

[91] Radhakrishnan, V. and Srinivasan, G. 1980, J. Astroph. & Astron. 1, 47

[92] Gomez-Gonzalez, J. and Guelin, M. 1974, A&A 32, 441

[93] Ables, J.G. and Manchester, R.N. 1976, A&A 50, 177

[94] Weisberg, J.M., Rankin, J., and Boriankoff, V. 1980, A&A 88, 84

[95] Reynolds, R.J. 1977, Ap.J. 216, 433

[96] Pottasch, S.R., Wesselius, P.R., and Van Duinen, R.J. 1979, A&A 74, L15

[97] Watson, W.D. 1972, Ap.J. 176, 103

[98]   - ibid -   , 271

[99] Jura, M. 1976, Ap.J. 204, 12

[100] de Jong, T. 1977, A&A 55, 137

[101] de Jong, T. 1980, Highlights of Astronomy 5, 301

[102] Spitzer, L. 1956, Ap.J. 124, 20

[103] Jenkins, E.B. 1978, Ap.J. 219, 855

[104] Jenkins, E.B. 1978, Ap.J. 220, 107

[105] Burnstein, P., Borken, R.J., Kraushaar, W.L., and Sanders, W.T. 1977, Ap.J. 213, 405

[106] Hayakawa, S., Kato, T., Nagase, F., Tanaka, Y., and Yamashita, K. 1978, A&A 62, 21

[107] Stern, R. and Bowyer, S. 1979, Ap.J. 230, 755

[108] Rocchia, R., Arnaud, M., Blondel, C., Cheron, C., Christy, J.C., Ducros, R., Koch, L., Rothenflug, R., Schnopper, H.W., and Delvaille, J.P. 1981, Sp. Sci. Rev. 30, 253

[109] Arnaud, M. Rocchia, R., Rothenflug, R., Soutoul, A. 1981 17th Int. Cosmic Ray Conf. 1, 131

[110] Rosner, R., Avni., Y., Bookbinder., J., Giacconi., R., Golub., L., Harnden., F.R., Maxson., C.W., Topka., K., and Variana, G.S. 1981 Ap].J.Lett. 249, L5

[111] McKee, C.F. and Ostriker, J., 1977 Ap.J. 218, 148

[112] Cox, D.P. 1979 Ap.J. 234, 863

[113] Cox, D.P. 1981 Ap.J. 245, 534

[114] McCray, R., and Snow, T.P., 1979 Ann. Rev. Astr. Ap. 17, 213

[115] McKee, C.F. 1982 in Supernovae, Rees and Stoneham (eds) Reidel (Dordrecht)

[116] Cox, D.P., and Anderson, P.R., 1982 Ap.J. 253, 268

[117] Habe, A., Ikeuchi, S., Tanaka, Y., 1981 Publ. Astr. Soc. Japan 33, 23

[118] Brand, P.W.J.L., and Heathcote, S.R. 1982 M.N.R.A.S. 198, 545

OPTICAL AND IR EMISSION LINES

M. Perinotto

Astrophysical Observatory of Arcetri, Florence, Italy

INTRODUCTION

   To discuss the present understanding of optical and IR
emission lines in the interstellar medium, we summarize the basic
theory involved, and next see the status of observations and the
approximations used to derive the physical parameters.

   Our goal is finally oriented toward the problem of chemical
gradients in galaxies as it can be studied from analysis of HII
regions and planetary nebulae.

ELEMENTS OF BASIC THEORY

1. THE PHYSICAL REGIME

   The physical regime in which we are interested includes particle densities of ~ 10 to $\sim 10^6$ cm^{-3} and temperatures of a few
thousands of degrees. We are thus dealing with a relatively dilute gas, where there are often very large departures from
thermodynamic equilibrium. We assume that the age of our objects
is large enough and bulk motions relatively unimportant so that
population of levels is governed by steady state equations and
the energy balance does not include terms due to shocks, nor to
volume variations. This is known to be the case for planetary
nebulae and HII regions. Other complications, as magnetic fields,
are neglected.

J. Audouze et al. (eds.), Diffuse Matter in Galaxies, 205–237.
Copyright © 1983 by D. Reidel Publishing Company.

2. STEADY STATE EQUATIONS

2.1 Basic formulation

The steady state equation for the number density N_i^s of the level with energy E_i of an ion (s-1) times ionized reads:

$$N_i^s \left\{ \sum_k \left[\overset{①}{A_{ik}} + \overset{②}{B_{ik} \, \bar{I}_{ik}} + \overset{③}{C_{ik} N_e} \right] + \sum_j \left[\overset{④}{B_{ij} \, \bar{I}_{ij}} + \overset{⑤}{C_{ij} N_e} \right] + \right.$$

$$\left. + \overset{⑥}{N_e C_{ic}} + \overset{⑦}{\int_{\nu_i}^{\infty} \frac{4\pi J_\nu}{h\nu} a_{ic,\nu} \, d\nu} \right\} = N_1^{s+1} N_e \left[\overset{⑧}{\alpha_i^s} + \overset{⑨}{N_e \phi_i^s} \right] + \quad (1)$$

$$+ \sum_j \overset{⑩}{N_j^s} (\overset{⑩}{A_{ji}} + \overset{⑪}{B_{ji} \, \bar{I}_{ji}} + \overset{⑫}{C_{ji} N_e}) + \sum_k \overset{⑬}{N_k^s} (\overset{⑬}{B_{ki} \, \bar{I}_{ki}} + \overset{⑭}{C_{ki} N_e})$$

where the energy of levels follows $E_k < E_i < E_j$; s denotes the state of ionization, s = 1 meaning neutral atom. The left hand side describes depopulation of level i via line radiation (processes 1, 2, 4), collisions with electrons toward bound states (3, 5) or the continuum (6) and radiative ionizations (7). The right hand side describes population of level i from two-body recombinations (radiative and dielectronic) (8), three-body recombinations (9), radiative processes from higher levels (10, 11) or lower levels (13) and collisions with electrons (12, 14). In the equation (1):

- A_{ik} $[s^{-1}]$, B_{ik} $[cm^2$, sterad $erg^{-1}]$, B_{ki} [idem] are the Einstein coefficients for spontaneous emission, stimulated emission and absorption, respectively. They obey:

$$B_{ik} \, g_i = B_{ki} \, g_k \quad ; \quad A_{ik} = \frac{2h\nu^3}{c^2} B_{ik} \quad (2)$$

The A_{ik} values range from 10 for permitted lines to $1-10^{-4}$ for most optical forbidden lines, $10^{-3} - 10^{-6}$ for several IR forbidden lines and goes down to $2.9 \, 10^{-15} \, sec^{-1}$ for the 21 cm line.

- C_{ik} , C_{ki} $[cm^3 \, s^{-1}]$ are the coefficients for superelastic deexcitation (no emission of photon) and inelastic excitation by collisions with electrons, respectively. They are evaluated from:

$$C_{ik} = \frac{8.629 \, 10^{-6} \bar{\Omega}(k,i)}{g_i \, T^{1/2}} \quad ; \quad C_{ki} = \frac{g_i}{g_k} C_{ik} \, e^{-E_{ik}/kT} \quad (3)$$

$$\text{with} \quad E_{ik} > 0$$

where $\bar{\Omega}(k,i)$ are the so called collision strengths averaged over a Maxwellian distribution of impact velocities at the temperature T.

At $T = 10^4$ K, since $\bar{\Omega} \sim 1$, we have

$$C_{ik} \simeq \frac{10^{-7}}{g_i} \quad ; \quad C_{ki} \simeq \frac{3 \ 10^{-8}}{g_k} \ e^{-(E_{ik})eV} \tag{4}$$

and in our density range $(N_e \simeq 10 - 10^6)$ it is:

$$A_{ik} > C_{ik} \ N_e \sim 10^{-7} \ N_e \tag{5}$$

for permitted lines but it is not so in general for forbidden lines. Notice also that at λ 1500 A, 5000 A, 10 μm, $e^{-h\nu/T}$ amounts to 6.8 10^{-5}, 5.5 10^{-2}, 0.87   illustrating how low lying levels are favorite for collisional excitation.

- C_{ic} [cm^3 s^{-1}]is the ionization coefficient for collision with electrons. The importance of C_{ic} relative to the radiative ionization (R_{ic}) can be evaluated as:

$$\frac{\text{term} \ \textcircled{7}}{\text{term} \ \textcircled{6}} = \frac{R_{ic}}{N_e C_{ic}} \approx 0.118 \ \frac{\nu_i^3}{c^3} \ (\frac{WT_R}{N_e T_e^{1/2}}) \exp \left[\frac{h\nu_i}{k} \ (\frac{1}{T_e} - \frac{1}{T_r}) \right] \tag{6}$$

where T_r is the radiation temperature and W the dilution coefficient. We find actually that the interstellar radiation field has approximately a black-body spectrum, appropriate to that of the exciting star but greatly diluted by the star's surface covering a very small solid angle. Thus, we write

$$I_\nu = W \ B_\nu \ (T_*) \tag{7}$$

where $T_* = 35\,000 - 50\,000$ °K   for an HII region and up to perhaps 300 000 °K for a planetary nebula and W may be of the order of 10^{-14} in both types of objects. Equation (6) is valid in the hydrogenic approximation for the photoionization cross-section a_{ic} , and in the limit that $h\nu_i \gg kT_r$ and $h\nu_i \gg kT_e$. With $T_r = T_{eff}$ of the exciting star, say 40 000 K, $T_e = 10^4$, $h\nu_i = 13.6$ eV, equation (6) gives

$$\frac{R_{ic}}{N_e \ C_{ic}} \simeq 2.4 \ 10^{22} \ \frac{W}{N_e} \tag{8}$$

If $W \simeq 10^{-14}$ and $N_e < 10^7$ cm^{-3}, the collisional ionizations are less efficient compared with the radiative ionizations. Thus the term in C_{ic} can be usually neglected except for levels of high principal quantum number (as hydrogenic states with $n \gtrsim 40$), which are responsible, in case of hydrogen, of the observed hydrogen radio recombination lines.

- α_i^S [cm^3 s^{-1}], ϕ_i^S [cm^6 s^{-1}] are the two bodies (radiative plus dielectronic) and three bodies effective recombination coefficients, respectively. In the first coefficient the radiative term usually prevails at our temperatures (but there are exceptions), is higher for the lowest levels where it takes typical values (for hydrogen) around $.10^{-14}$ cm^3 s^{-1} and decreases slowly with the temperature. The dielectronic term becomes important at temperatures higher than 10^5 K. The second coefficient can be expressed (omitting overscripts) as:

$$\phi_i = i^2 \left(\frac{h^2}{2\pi m\, k\, T} \right) e^{h\nu_i/kT} \cdot C_{ic} \tag{9}$$

C_{ic} can be approximated with:

$$C_{ic} = 1.55 \ 10^{13} \ T^{-1/2} \ \bar{g}_i \ \frac{\alpha(\nu_i)\exp(-h\nu_i/kT)}{h\nu_i/kT} \tag{10}$$

with $\bar{g}_i \simeq 0.1$ for hydrogenic ions, $\alpha(\nu_i)$ the threshold photoionization cross-section.
Thus it is

$$\phi_i \simeq 6.43 \ 10^{-4} \ i^2 \ \alpha(\nu_i) \ \frac{1}{T^2} \ \frac{k\,T}{h\nu_i} \tag{11}$$

and for hydrogen, $T^4 = 10$, $i = 40$

$$N_e \ \phi_i \simeq 2.6 \ 10^{-24} \left(\frac{k\,T}{h\nu_i} \right) N_e = 2.7 \ 10^{-22} N_e \tag{12}$$

$$\simeq 2.7 \ 10^{-16} \ \text{for} \ N_e = 10^6 \ \text{cm}^{-3} \quad .$$

This value should be compared with α_{40} . For H the recombinations coefficients are (n $\equiv$ i):

$$\alpha_n = 2.5 \ 10^4 \ \frac{g_{nf}}{n^3 T^{1/2}} \left(\frac{1}{3.12 \ 10^{14} + \dfrac{3.29 \ 10^{15}}{n^2}} \right) \tag{13}$$

where g_{nf} is the Gaunt factor for the bound-free transition. For n = 40, T = 10^4 , it is

$$\alpha_{40} \simeq 1.2 \ 10^{-17} \ g_{nf}$$

Thus for levels with high values of the principal quantum number (responsible in H for the hydrogen radio recombination lines) the three-body recombinations become important.

2.2 Additional terms in equations (1)

Excitations to particular levels may be contributed:

1E) by collisions with neutral atoms (as in hydrogen with the excitation from the ground 1^2S term to 2^2P, followed by emission of an L_α photon, and to 2^2S, followed by the two

2E) by absorption of a photon produced by a different atom in almost coincidence of wavelength (excitation by fluorescence as in the named Bowen mechanism involving the He II Ly_α 303.78 line and the OIII $2p^2$ $^3P_2$ - 3d $^3P^\circ_2$ line at λ 303.80).

Ionizations, in addition to those produced by direct radiation from the ionizing source, may occur:

1I) from diffuse photons produced by recombinations within the nebula. Important are the high density photons of resonance lines, in particular of HI, HeII. Also the continuous diffuse radiation field at frequencies just higher than the ionization edges is fairly important. All the corresponding terms enter in the expression of J_ν ;

2I) via collisions among ions of heavy elements and neutral hydrogen (or, but with rates apparently very small, with helium) through the so-called charge transfer processes. These processes have been recognized to be important in particular to explain the relatively high intensity of the optical forbidden lines of the low ionization ions OII, NII (e.g. Perinotto, 1977; Pequignot, 1978).

3. IONIZATION EQUATION

By adding equations (1) for different levels, the terms connecting bound states cancel out and one gets:

$$\sum_{i=1}^{M} N_i^s \left[N_e C_{ic} + \int_{\nu_i}^{\infty} \frac{4\pi J_\nu}{h\nu} a_{ic,\nu} \, d\nu \right] = \sum_{i=1}^{M} N_1^{s+1} N_e \left[\alpha_i^s + N_e \phi_i^s \right] \quad (14)$$

where M is the number of levels considered.

Equation (14) is called the ionization equation. The electron density clearly comes from:

$$N_e = \sum_{A} \sum_{s=2}^{M_A} (s-1) N_{As} \quad (15)$$

where A denotes the atomic species and s the state of ionization.

Almost all electrons come from H and He in ionized nebulae so that there we have

$$N_e \simeq 1.1 \ N_H \quad (16)$$

Instead electrons come mainly from C^+ and Si^+ in HI regions, where the radiation field of stars can ionize C° and Si°.

4. THERMAL BALANCE

The thermal balance equates heating to cooling, and writes:

(Energy gained by gas) = (Energy losses from recombination) +
(Energy losses from collisional excitation of levels).

Specifically it can be expressed (Harrington, Seaton, Adams and Lutz, 1982):

$$\sum_A \sum_{s=1}^{M_A-1} N_{As} \psi_{As} = \sum_A \sum_{s=2}^{M_A} N_{As} (\beta_s^{ff} + \beta_{As}^{bf}) N_e +$$

$$+ \sum_A \sum_{s=1}^{M_A} N_{As} \left[\sum_{u=2}^{L_{As}} P_{As}(u) \sum_{\ell=1}^{u-1} A_{u\ell} h\nu_{u\ell} \right];$$

(17)

with

$$\psi_{As} = \int_{\nu_{As}}^{\infty} \frac{4\pi J_\nu}{h\nu} a_{As}(\nu) (h\nu - h\nu_{As}) d\nu$$

Here ψ_{As} is the rate of energy input by the photo-ionization per ion of species A and state of ionization s, β_s^{ff} is the free-free energy loss, β_{As}^{bf} is the energy lost by recombining electrons and the last term represents the cooling when the low-lying energy levels of the ion (of which the number considered is L_{As}) are collisionally excited and decay by emission of radiation. In this expression $P_{As}(u)$ is the relative population of the upper level $u(\sum_u P_{As}(u) = 1)$ and ℓ denotes the lower level of the transition , and follows from the solution of the relevant steady state equations (1).

In evaluating the cooling term, all levels of abundant heavy ions within about 10 eV of the ground state should be included, along with the $u = 2$ and $u = 3$ levels of hydrogen.

5. RADIATIVE TRANSFER

Equations (1) are coupled with equations of radiative transfer since one needs to specify quantities as I_{ik} and J_ν appearing in (1). The radiative transfer equation involving I_ν is:

$$\frac{dI_\nu}{dx} = - \kappa_\nu \rho I_\nu + \varepsilon_\nu \rho/4\pi \quad .$$

(18)

5.1 The lines

For the lines, one writes (i > k):

$$\Delta\nu \, \frac{dI_{ik}^{o}}{dx} = - \left[(N_k B_{ki} - N_i B_{ik}) I_{ik}^{o} + N_i A_{ik} \right] \frac{h\nu_{ik}}{4\pi} \tag{19}$$

The optical depth in the line is

$$\Delta\nu \, \frac{d\tau_{ik}^{o}}{dx} = N_k (B_{ki} - \frac{N_i}{N_k} B_{ik}) \frac{h\nu_{ik}}{4\pi} \tag{20}$$

where $\Delta\nu \cdot I_0$ is proportional to the total energy in the line (i,k) with central intensity I_0 and full width at half maximum intensity $\Delta\nu$. Writing:

$$\frac{N_i}{N_k} = \frac{g_i}{g_k} \frac{b_i}{b_k} e^{-h\nu_{ik}/kT} \tag{21}$$

with g_i, g_k statistical weights and b_i, b_k coefficients measuring the deviation from LTE of population of the two states, it is for the optical depth per unit frequency at the line centre τ_{ik}^{o}, with x the geometrical depth:

$$\frac{d\tau_{ik}^{o}}{dx} = N_k \frac{g_i}{g_k} \frac{\lambda_{ik}^{2}}{8\pi^{3/2} \Delta\lambda_D} A_{ik} (1 - \frac{b_i}{b_k} e^{-h\nu_{ik}/kT}) \tag{22}$$

or even, in case of $h\nu_{ik}/kT \ll 1$,

$$\frac{d\tau_{ik}^{o}}{dx} = N_k \frac{g_i}{g_k} \frac{\lambda_{ik}^{2}}{8\pi^{3/2} \Delta\nu_D} A_{ik} (\frac{b_i}{b_k} \frac{h\nu_{ik}}{kT} - \frac{d \ln b_k}{dk} \Delta k) \tag{23}$$

where $\Delta k = i-k$. We are referring to lines broadened by thermal Doppler effect. Thus

$$\Delta\nu_D = (\frac{2kT}{Mc^2})^{1/2} \nu_o \tag{24}$$

With M the mass of the atom.

The factor in parenthesis in eqs. (22) and (23) represents the correction for stimulated emission. It is negligible for optical and UV lines, ($h\nu \gg kT$), but important for lines whose energy is smaller than kT, i.e. at T = 10^4 K, $\lambda > 1.4$ μm.

If $h\nu \gg kT$, it is (from eq.(21)) and considering that the lowest levels deviate more from LTE,

$$N_i \, B_{ik} << N_k \, B_{ki} \tag{25}$$

Thus eq. (19) simplifies. If furthermore the absorption is small,

$$N_k \, B_{ki} \, I_{ki} << N_i \, A_{ik} \tag{26}$$

Eq.(19) is then simply integrated giving the intensity emerging from the column

$$I_{ik} = \frac{h\nu_{ik}}{4\pi} \int N_i \, A_{ik} \, dx \tag{27}$$

If $h\nu < kT$, we have with increasing λ two opposite effects. The absorption coefficient, uncorrected for stimulated radiation, increases with λ^3 often overbalancing the reduction of the A_{ik} values. In fact, for instance, A_{ik} become smaller by $\sim 10^2 \div 10^3$ going from optical forbidden lines to IR forbidden lines involving the ground configuration. E.g. it is for [OIII] 52 μm/5007 A :

$$\frac{A(^3P_2 - ^3P_1)}{A(^1D_2 - ^3P_2)} \, \frac{\lambda^3(^3P_2 - ^3P_1)}{\lambda^3(^1D_2 - ^3P_2)} = \frac{9.8 \; 10^{-5}}{2.1 \; 10^{-2}} \left(\frac{52 \; \mu m}{5007 \; A}\right)^3 = 5.2 \; 10^3 \tag{28}$$

The corresponding factor for stimulated emission instead decreases toward $\left(1 - \frac{b_i}{b_k}\right)$ and may become negative if $b_i > b_k$ concurs with $h\nu/kT$ small enough.

5:2 The continuum

A problem of transfer is to be considered also for the ionizing continuum, since its optical depth is large.

6. THE OVERALL PROBLEM

The steady state equations (1), the thermal balance equations (17) and the radiative transfer equation (18) are to be solved to obtain the energy emitted by a model nebula at all wavelengths.

The distribution of nebular matter N_H , usually spherical symmetric around a central source, its ionizing flux and the chemical abundances in the nebula must be specified, as inputs to the model.

The emergent radiation, in particular in the lines, is so computed and compared with the observations. Some of the mentioned inputs, as the chemical abundances, are taken as parameters

and then derived from the best fit of all observed lines with
computed values. This is in principle the best method to obtain
chemical abundances in nebulae.

7. CALCULATION OF LINE SPECTRUM

7.1 Hydrogen

If the nebula is <u>optically thin</u> in all the lines (case A),
the radiative transfer problem in the lines is neglected. One
solves the (M-1) equations $i \geq 2$ of system (1) to get the relative populations of hydrogen levels as function of T_e , N_e ,
$N(H^+)$. The ionization equation (14), giving $N(H^+)/N(H^\circ)$, is
coupled with analogous equations connecting He° to He^+ and He^+
to He^{++} via the diffuse photons mentioned under item 2.2.2I.

The important terms in (1) for levels responsible of <u>optical</u>
lines are 1, 8, 10. As a consequence the emitted lines are called
"recombination lines". The emissivity is expressed as

$$j_{nn'} = \frac{h\nu_{nn'}}{4\pi} \sum_{L=0}^{n-1} \sum_{L'=L\pm1} N_{nL} A_{nL,n'L'} = \frac{h\nu_{nn'}}{4\pi} N_p N_e \alpha^{eff}_{nn'}(T_e,N_e) \quad (29)$$

where n,L denote principal and azimuthal quantum numbers respectively. In particular for H_β the intensity from a particular
direction is

$$I(\lambda 4861) = \frac{h\nu_{4861}}{4\pi} \int N_p N_e \alpha^{eff}_{H_\beta} (H^\circ,T) \, dx \quad . \quad (30)$$

If the nebula is (more realistically) <u>optically thick in the</u>
Lyman lines (case B), one assumes that each emitted Lyman photon
is locally reabsorbed. Then the solution of system (1) proceeds
in a similar way except now index k remains ≥ 2.

For the so called <u>radio hydrogen recombination lines</u>, the
important terms, in addition to (1,8,10), are the collisional
terms (3,5,12,14), the collisional ionization (6) and its inverse
process (9). The solution may be expressed in terms of the
coefficients b_n giving the deviation from LTE population and,
in case B, the emergent intensity is evaluated from the transfer
equation (19) using (23).

7.2 Helium

The spectrum of He I singlets is very similar to that of HI.
He I Lyman lines are able to ionize H°, as already mentioned. The
spectrum of He I triplets is modified by the presence of the hi-

ghy metastable 2^3S level that is strongly populated, so that self absorptions from it in the $2^3S - n^3P$ series (λ 10830, 3889 etc. with the consequences for the enhanced cascades from the upper level of λ 3889, etc.), collisional excitation to the upper level of λ 10830, and collisional deactivation of it toward 2^1S and 2^1P levels are important. The series $2^3P - n^3D$ (λ 5876, 4471 etc.) is less affected by these various processes.

The spectrum of He II is evaluated as the one of H I. The He II Ly_α line at λ 303.78 produces however the fluorescence excitation of O III lines mentioned under (2.2.2E). This process clearly does not affect significantly the helium spectrum. The He II Ly_α line as the other He II Lyman lines can ionized He°, H° so coupling the ionization equations of H°, He° and He⁺.

The best observed optical lines of He I and He II are λ 5876 and λ 4686 respectively and these are evaluated from

$$I(\lambda 5876) = \frac{h\nu_{5876}}{4\pi} \int N(He^+) \, N_e \, \alpha^{eff}_{5876}(He°,T) \quad dx \qquad (31)$$

$$I(\lambda 4686) = \frac{h\nu_{4686}}{4\pi} \int N(He^{++}) N_e \, \alpha^{eff}_{4686}(He^+,T) \quad dx \qquad (32)$$

7.3 Heavy elements

Due to low abundances relative to H and He, the recombination spectrum of heavy elements is very weak. The circumstance however occurs that the ground configuration displays often from two to five levels within few eV from the lowest energy state, and a few n > 1 levels are also close to the ground state. All these levels are low enough to be effectively excited by collisions with electrons. Thus collisional excited permitted (mostly in the UV), forbidden and intercombination lines (from IR to the UV) are produced. As with the metastable levels, deactivation is not frequent enough because of the low density of matter and radiation, while the excitation is frequent enough so that some of the forbidden radiation may result stronger than recombination lines of hydrogen.

Equations (1) reduce to the collisional terms 3,5,12,14 and to the spontaneous radiative ones 1, 10. The solution of (1) are evidently fairly sensitive to T_e , N_e :

$$\frac{N_i}{N_1} = f \, (T_e \, , \, N_e) \qquad (33)$$

so that line intensity ratios of radiations of the same ions offer excellent tools to determine T_e , N_e across the nebula. The emergent intensity in these forbidden lines can be written

$$I_{ik} = \frac{h\nu_{ik}}{4\pi} A_{ik} \int N_i \, dx = \frac{h\nu_{ik}}{4\pi} A_{ik} \int N_1 \, f(T_e, N_e) \, dx \qquad (34)$$

For collisionally excited lines, as far as the collisions from lowest level is the dominant populating mechanism, one may also write:

$$I_{il} = \frac{h\nu_{il}}{4\pi} \int N_1 \, N_e \, \frac{8.63 \ 10^{-6}}{T^{1/2}} \, \frac{\Omega(1,i)}{g_1} \, e^{-\chi_{il}/kT} \, b_i \, dx \qquad (35)$$

with b_i expressing the fraction of ions of level i that undergoes a radiative transition to the ground state. Eq. (35) holds for optically thin lines.

8. ROLE OF DUST

Dust is mixed more or less homogeneously with gas in HII regions and, less heavily, in typical planetaries. The dust absorbs particularly high density radiation, as ionizing continuum (but even optical continuum photons) and resonance line photons. The absorbed energy is reemitted at the dust equilibrium temperature, i.e. in the near and far IR.

As a result, in presence of dust, the size of the various ionization zones become smaller, and ionic abundances deduced from resonance lines are underestimated, if not properly corrected. Moreover dust scatters light, so that light observed in a particular zone may contain contribution of light produced in different zones (see e.g. Mathis et al., 1981).

The real importance of these effects depends on the absorption and scattering cross-section of the dust. These are unknown in the UV range ($\lambda < 912$ A).

OBSERVATIONS OF LINE EMISSION IN PHOTOIONIZED NEBULAE AND EVALUATION OF PHYSICAL AND CHEMICAL PROPERTIES

9. THE OBSERVED EMISSION LINES

9.1 Planetary nebulae

A large body of observations of individual lines of planetary nebulae (and HII regions) is collected in the compilation of Kaler (1976). For subsequent work see references in IAU Symposia

N° 76 (1978) and N° 103 (1982). NGC 7027 has probably the ri-
chest observed speçtrum with ∿500 measured emission lines from
λ 3047 to λ 10938 Å. The IUE satellite has added some 20 lines
in the spectral range 1150 to 3150 A (cf. Harrington et al.,
1982). IR lines of various elements have been observed (cf.
Simpson, 1975; Beck et al., 1981; Dinerstein, 1983). The responsible ions, from the IR to the UV range, are shown in Table 1.

9.2 HII regions

        The best observed HII region is the Orion nebula. Some 233
lines have been observed from λ 3187 to λ 10938 (Aller and Liller,
1959; Kaler et al., 1965; Peimbert and Torres-Peimbert, 1977).
Few additions come from the UV (Perinotto and Patriarchi, 1980)
and the IR spectral range (cf. Simpson, 1975). The line spectrum
of an HII region resembles that of a low excitation planetary nebula.

10. NATURE OF THE OBSERVED EMISSION LINES

        The observed emission lines are interpreted as:

   a) Recombination permitted lines of H and He (as Lyman, Balmer,
      Paschen etc. series of H; the $2^3P - n^3D$ series of He I:
      λ 5876, 4471 etc; the He II Balmer series:λ 1640, 4686 etc.)
      and weak lines of heavy elements.

   b) Collisionally excited permitted lines of heavy elements (es-
      sentially resonance lines in the UV, as CIV 2s ^2P - 2p ^2P
      λ 1548, 1550; SiIV 3s ^2S - 3p ^2P λ 1394, 1403).

   c) Collisionally excited forbidden lines of heavy elements
      (from the IR to the UV spectral range).

   d) Resonance-fluorescence excited permitted lines (as those in
      cascade from the 3d $^3P_2^o$ level of OIII excited from the
      λ 303. 78 HeII, as 3133, 3340, 3444 etc.).

11. CRITERIA FOR DIAGNOSTIC OF PHYSICAL CONDITIONS

        It has been recalled that lines reported under 10c provide
valuable criteria of diagnostic of physical conditions in the regions where the radiation forms. A number of these criteria in
the optical range is collected in Table 1 of Perinotto (1971).
Other ratios can be formed using fine structure forbidden lines,
falling in the IR spectral range (cf. Simpson, 1975), or intercombination lines.

TABLE 1.   IONS PRODUCING OBSERVED EMISSION LINES FROM UV TO IR IN PLANETARY NEBULAE (NGC 7027, 7662) AND IN THE ORION NEBULA.   THE IONS SEEN ONLY IN PLANETARY NEBULAE ARE UNDERLINED.

H	I
He	I, II
C	<u>I</u>; II, II]; <u>III, III</u>]; <u>IV</u>
N	[I]; II,[II]; III, <u>III</u>],[III]; <u>IV</u>, IV]; <u>V</u>
O	<u>I</u>,[I]; II,[II]; <u>III, III</u>],[III]; <u>IV, IV</u>]; <u>V</u> ; <u>VI</u>
F	[<u>IV</u>]
Ne	II,[II];[III];[<u>IV</u>];[<u>V</u>]
Na	[<u>IV</u>];[<u>V</u>]
Mg	I,[I]; II;[<u>IV</u>];[<u>V</u>]
Si	II, <u>III, III</u>],[<u>SiIV</u>]
P	<u>II</u>
S	[<u>I</u>];[II]; III,[III];[<u>IV</u>]
Cℓ	[II];[III];[<u>IV</u>]
Ar	[<u>II</u>],[III];[IV];[<u>V</u>]
K	[<u>IV</u>];[<u>V</u>];[<u>VI</u>]
Ca	[<u>V</u>];[<u>VII</u>]
Mn	[<u>V</u>];[<u>VI</u>]
Fe	[II];[III];[<u>IV</u>];[<u>V</u>];[<u>VI</u>];[<u>VII</u>]

In cases, particularly in extragalactic HII regions, no one
of the above mentioned criteria, in particular for determining
T_e , can be used because of faintness of the relevant lines. Then
ratios involving lines of different atoms have been proposed, as
$\{\lambda(4959+5007)[OIII]+ \lambda(3727+29)[OII]\}/H_\beta$ (Pagel et al., 1979) and
$\lambda(4959+5007)[OIII]/\lambda(6548+6584)[NII]$ (Alloin et al., 1979), and
empirical relationships of these ratios versus T_e have been
constructed. The problems with the use of these ratios are however not negligible (Stasinska et al., 1981).

12. METHOD TO DETERMINE CHEMICAL ABUNDANCES

To determine chemical abundances in photoionized nebulae one
may distinguish between detailed models, simplified procedures
and intermediate procedures.

12.1 Detailed models

This procedure (see e.g. Harrington et al., 1982) requires
the calculations of the emitted spectrum from a model nebula in-
tended to represent the observed object. One must specify:
a) the radiation field of the exciting star, b) the geometry of
the nebula and its mass distribution, c) the chemical composition of the nebula taken to be constant through it, d) the dust
content and its properties.

Essentially all the physical concepts described from Section
1 to 8 are to be applied. The emission spectrum is calculated
and compared with the observations. The chemical abundances are
taken as a parameter and deduced from this comparison. This procedure is clearly the most satisfactory one in principle, but it
is very complicated, requires huge of time computing and suffers
from uncertainties in the input quantities. That is why only very few objects have been studied with this method so far: essentially only NGC 7662 (Harrington et al., 1982) and IC 3568
(Harrington and Feibelman, 1982). The method seems anyhow best
suited for planetary nebulae because of their more regular geometry and less concentration of dust, than for HII regions.

We mention as the best example of this method the study of
NGC 7662 by Harrington et al. (1982). The agreement between calculated and observed spectrum is good for all the lines well measured, with the apparent exception of $\lambda 3426$ [Ne V] which is computed to be three times stronger than observed (cf. Table 19 of
that work).

Since the agreement is often better than by a 30%, one infers
that the corresponding chemical abundances may be as accurate as that.

12.2 Symplified procedures

With these procedures one uses as many as possible criteria to determine T_e and N_e across the nebula and from these quantities applied to the various ions, obtains the chemical abundances of observed ionic species, from equations similar to (35).

Then one uses appropriate expressions to evaluate the contribution of unseen ions. These formulas have been constructed essentially on the basis of the ionization potential of the various ions, on the hypothesis that the relative concentration of the various ions follows their ionization potential. The most widely used of these expressions have been proposed by Peimbert and Costero (1969). They make wide reference to 0, that is commonly observed as O^+, O^{++}, and read:

$$\frac{N(O)}{N(H)} = \frac{N(He^+ + He^{++})}{N(He^+)} \frac{N(O^+ + O^{++})}{N(H^+)} \tag{36}$$

$$\frac{N(N)}{N(H)} = \frac{N(O)}{N(O^+)} \frac{N(N^+)}{N(H^+)} \tag{37}$$

$$\frac{N(Ne)}{N(H)} = \frac{N(O)}{N(O^{++})} \frac{N(Ne^{++})}{N(H^+)} \tag{38a}$$

$$\frac{N(S)}{N(H)} = \frac{N(O)}{N(O^+)} \frac{N(S^+ + S^{++})}{N(H^+)} \tag{39a}$$

An empirical correction formula to account for the presence of He° (Peimbert and Costero, 1969) is

$$\frac{N(He)}{N(H)} = \frac{N(He^+)}{N(H^+)} \left\{ 0.13 \frac{O}{O^{++}} + 0.87 \left[S/(S-S^+) \right] \right\} \tag{40a}$$

or even

$$\frac{N(He° + He^+)}{N(H)} = \frac{1}{\left[1-\gamma \frac{O^+}{O} - (1-\gamma) \frac{S^+}{S} \right]} \frac{He^+}{H^+} \tag{40b}$$

(Peimbert and Torres-Peimbert, 1977). They are based on the hypothesis (reflecting the ionization potentials):

$$\frac{S^+}{S} \leq \frac{He°}{H} \leq \frac{O^+}{O}$$

where γ is to be adjusted to reduce the dispersion from observation in various nebular positions and is found in the Orion nebula to be $\gamma \simeq 0.35$.

The validity of these ionization correction formulas (ICF) and of the division of the nebula in temperature zones, i.e. of the simplified procedures, can be studied with models of nebulae or tested observationally.

12.2a Analysis with models and improvements of ionization correction.

Studies of this kind have been made e.g. by Balick and Sneden (1976), Grandi and Hawley (1978), French and Grandi (1981) and Stasinska (1978,1980), with models adequate to represent HII regions or low excitation planetary nebulae. According to French and Grandi (homogeneous models with $N_e = 100$ cm^{-3}) the simplified procedures give abundances of N, O, Ne, S (using equations 38,39, 40a,41a) within a 50%. The analysis of Stasinska appears more valuable since she uses more modern values for some atomic quantities and a better treatment of the diffuse radiation field. Anyhow differences might be expected if one adopts different models for the ionizing radiation of the exciting star from the Mihalas (1972) NLTE unblanketed atmospheres chosen by Stasinska (see Lucy et al., 1982). Certainly these differences will be clarified by future work. By now we rely here on the mentioned works of Stasinska.

Stasinska: i) confirms eq. 37 (for N), ii) notes that eq. 40b overestimates He° if $O^{++}/O < 0.5$ (what may apply to galactic HII regions and to a few Magellanic Clouds objects), iii) criticizes eq. 39a (for S) as overestimating S_{tot} by as much as a factor of ~ 10 for $T_* \gtrsim 40\,000$ K and iv) proposes:

$$\frac{Ne^{++}}{Ne} = 1.2 \frac{O^{++}}{O} - 0.2 \quad (\text{if } \frac{O^{++}}{O} > 0.2) \tag{38b}$$

$$\frac{Ne}{H} = \frac{Ne^{++}}{H^+} \left[\frac{O^+ + O^{++}}{O^{++} - 0.2\,O^+} \right]$$

$$\frac{S}{H} = \left\{ 1 - \left[1 - \frac{O^+}{O} \right]^3 \right\}^{-1/3} \left(\frac{S^+ + S^{++}}{H^+} \right) \tag{39b}$$

If one observes the [SIII] 9069, 9532 lines; otherwise:

$$\frac{S^+}{S} = \frac{O^+}{O} \quad \text{for hot stars} \quad (\frac{4959+5007}{3727} \gtrsim 10 \text{)}$$

$$\frac{S}{S} = (\frac{O^+}{O})^2/2.5 \quad \text{cor colder stars.}$$

(39c)

She estimates that for $Z \lesssim Z_o$ the simplified procedures, with the expressions above suggested by her, gives abundances for N, O, Ne within 20-30% and within a factor of 2 or a 30% for S according to availability of the λ 9069+9532 lines. All this refers to adopting everywhere in the nebula the electron temperature from [OIII] 4363/5007 and to homogeneous objects. If $Z > Z_o$, the procedure may underestimate the true abundances by factors of 5-10 so giving the erroneus indication of Z lower than solar.

Equation (39b) is considered to be valid for PN too, from empirical tests (see next section), as well as equations 36, 37, 38a (for O, N, Ne) by Barker (1980) in his study of the planetary nebula NGC 6720. Moreover Barker (1980) suggests for argon in PN:

$$\frac{Ar}{H} = (\frac{S^+ + S^{++}}{S^{++}}) (\frac{Ar^{2+} + Ar^{3+} + Ar^{4+}}{H^+})$$

(41a)

that is however believed to underestimate Ar in zones of low excitation $(O^+ > O^{++})$ and consequently in some HII regions.

Another expression considered by Barker valid within a factor of 2 in most PN is

$$\frac{Ar}{H} = 1.5 \frac{Ar^{2+}}{H^+}$$

(41b)

French and Grandi (1981) suggest for argon in low ionization nebulae,

$$\frac{Ar}{H} = \frac{(Ar^{2+} + Ar^{3+})}{H^+} \frac{He}{He^+}$$

(41c)

and French (1981) for all nebulae from empirical test (see next Section):

$$\frac{Ar}{H} = (\frac{Ar^{2+} + A^{3+} + A^{4+}}{H^+}) (\frac{He}{He^+ + He^{++}})$$

(41d)

French (1981) also suggests that number $(3; -1/3)$ of eq. (39b) for S be substituted with $(2; -1/2)$.

12.2b Observational test of simplified procedures

The simplified procedures may be tested by applying them to observations in different points of a nebula and expecting iden-tical total abundances.

This has been done in planetary nebulae by Hawley and Miller (1977, 1978), Hawley (1978a), who from spectra in different posi-tions in NGC 6720, 6853, 7293 obtain variations by less than a factor of ~ 2 among the helium, oxygen, nitrogen and sulfur abun-dances derived with equations 40a, 36, 37, 39a, respectively, for the different positions in each nebula; the results for neon are less good. It has also been done by Barker (1980) (cf. Section 12.2a) and by French (1981) who studies a group of PN in the so-lar neighbourhood, assuming they have the same abundance in Ar.

Results of these observational tests indicate an accuracy of around a factor of 2, less than the one indicated by models analysis, and the fact may be ascribed to the model nebulae being always an idealization of the complexity (inhomogeneity, devia-tion from spherical simmetry, etc.) of real nebulae.

12.3 Intermediate procedures

These consist in reproducing with specific detailed models a number of relevant line ratios describing the excitation con-ditions, as

$$[OIII] / [OII] \quad , \quad HeII/HeI \quad , \quad [NeV] / [NeIII] \quad ,$$

then in obtaining ionic abundances as with simplified procedures, but correcting for unseen ions with the aid of the chosen model. This method seems quite promising. It has been extensively ap-plied e.g. by Aller and Czyzak (1982) to planetaries.

12.4 Comparison of methods

We compare in Table 2 results for NGC 7662 obtained by Harrington et al. (1982); Peimbert and Torres-Peimbert (1981); Benvenuti and Perinotto (1981), Aller and Czyzak (1982). The first study uses a highly sophisticated detailed model and all the available spectral information, the second the simplified procedure and optical spectrum, the third the simplified procedu-re and the UV spectrum only (1200 - 3100 A), the last the inter-mediate procedure with the whole spectral range.

Solar abundances are also shown by comparison. It is seen from Table 2 that the best agreement with values from the detai-

TABLE 2. ABUNDANCES IN NGC 7662 (GASEOUS COMPONENT)

	HSAL(1982)		PE(1971)	BP(1981)	AC(1982)	SUN
He	0.094	10.97	11.00	–	11.07	–
C	6.2-4	8.79	–	8.67	8.60	8.67
N	6.0-5	7.78	7.83	7.96	7.87	7.99
O	3.6-4	8.56	8.61	8.72	8.57	8.91
Ne	7.0-5	7.85	–	7.95	7.83	8.18
Mg	8.0-7	5.90	–	–	–	7.62
Si	6.0-6	6.78	–	–	–	7.63
S	1.5-5	7.18	–	–	6.83	7.23
C/O	1.7		–	0.89	1.1	0.57
N/O	0.17		0.17	0.17	0.20	0.12
Ne/O	0.19		–	0.17	0.18	0.18
O/O$_{Sun}$	0.43		–	0.63	0.46	–

led model is attained by the intermediate procedure of Aller and Czyzak. Differences are in general smaller than 0.1 dex. Other determinations in Table 2 are however within 0.2 dex, that is by a factor of 1.6, from the ones of the detailed model.

To explain the line emission of C , is to be outlined, one needs to consider the absorption by dust in the λ 1550 CIV line and moreover the actual geometry (if a sphere or a shell)(Koppen and Wehrse, 1982).

In conclusion the simplified procedures may be accurate to within a factor of two, the intermediate procedure appears to gi-give better results for O and Ne but not quite for C and S while the detailed model is expected to produce abundances accurate even to within a 30% (cf. Section 12.1). The best accuracy may be necessary for deciding for instance whether an object is carbon rich (C/O >1) or oxygen rich (cf. Table 2).

12.5 *Further analysis of simplified procedures*

Other comments on the limits of the simplified procedures are in order. This is particularly important for galactic and, the more, extragalactic HII regions where the available information on line intensities may be very scarce.

12.5a *Dust*

Effects of dust in HII regions have been studied by various authors, as Mathis (1970) who presents a way to discriminate internal from external dust, and Sarazin (1977). One must know the absorbing and scattering properties of dust, that may variate from place to place or from a galaxy to another, possibly in dependence of the content of heavy elements. For properties of dust see Savage and Mathis (1979), and Mathis (1978).

With an absorbing cross section constant with λ or increasing toward the UV, the main effects for the determination of chemical abundances of the gases, are:

 a) to reduce the total volume of ionized gas;
 b) to decrease the mean ionization degree of the elements;
 c) to reduce the Te gradient inside the nebula.

The three effects are in practice equivalent to a reduction of T_{eff} of the exciting star and thus are not much severe for their consequence on the derived abundances. On the other hand there are arguments (Mathis, 1982, private communication) for the dust absorption cross section to decrease toward the very far UV.

The dust scattering has effects recalled in Section 8, that are not easy to account for and so are generally neglected.

12.5b *Distribution of matter*

To see the effects due to variation in density and in other parameters, we report in Table 3 values from model calculations of Stasinska (1980).

In column 2 there are densities in an inner and an outer zone, in which the model consists, in column 3 the effective temperature of the exciting star, in column 4 the chemical abundance of heavy elements (Z = 1 stands for solar composition) in column 5 the intensity ratio λ (5007+4959) [OIII]/(6548+6584)[NII] (cf. Section 11 for its use in HII regions), in column 6 and 7 the electron temperatures from λ 5755/6584 [NII] and λ 4363/5007[OIII] respectively (T_r) and in column 8 the "line" electron temperature, relative to the λ (5007+4959) lines, T_ℓ(OIII), defined as:

$$T_\ell^{-1/2} \exp(-\Delta E/kT_e) = \frac{\int N(O^{2+})N_e^{-1/2} \exp(-\Delta E/kT_e)\ dV}{\int N(O^{2+})\ N_e\ dV} .$$

that is clearly best representative of the temperature to be used for deriving the abundance of O^{2+}.

It is seen from Table 3 that different densities in homogeneous nebulae (cases 1, 2, 3) modifie moderately the electron temperature and so the ratio [OIII]/[NII]. This ratio changes clearly in presence of density gradients across the nebula (case 4).

12.5c *Effective temperature and luminosity of exciting source*

The effective temperature has an important effect on the electron temperature in the nebula and therefore in the ratio [OIII]/[NII] (cases 5,6,7 of Table 3).

An increase by a factor of 100 of L_* (from model 11 to model 12 of Table 4) produces a steeper gradient of T_e with the consequence that adopting everywhere T_r (OIII) significantly overestimates abundances of low ionization species.

12.5d *Chemical abundances and electron temperature*

A decrease of Z (from $Z/Z_0 = 3$ to $Z/Z_0 = 0.2$) causes an increase of T_e by more than a factor of 2 and correspondin-

TABLE 3.   MODEL NEBULAE (Stasinska, 1980)

Model No.	$N(H)^a$		T_{eff}	Z	[OIII]/[NII]	$T_r(NII)$	$T_r(OIII)$	$T_\ell(OIII)$
	N_i	N_o						
1.	10	10	35 000	1.0	0.64	6840	6400	6310
2.	100	100	35 000	1.0	1.02	7040	6350	6290
3.	1000	1000	35 000	1.0	1.99	7300	6780	6780
4.	10	100	35 000	1.0	0.29	6940	6170	6110
5.	100	100	35 000	1.0	1.02	7040	6350	6290
6.	100	100	45 000	1.0	13.2	9190	7850	7700
7.	100	100	55 000	1.0	26.3	10150	8750	8560
8.	100	100	35 000	3.0	0.027	5230	4420	3320
9.	100	100	35 000	1.0	1.02	7040	6350	6290
10.	100	100	35 000	0.2	7.26	11100	11600	11600
11.[b]	100		35 000	1.0	1.00[c]	7200	6440	6262
12.[b]	100		35 000	1.0	0.74[c]	7700	6300	6048

a. Models are divided in two zones with different N_H(inner and outer).

b. Models from Stasinska (1978) with L_*(mod 12) = 100 L_*(mod 11).

c. These are only relative values.

gly an increase in emissivity of the [OIII] lines/H_β (at conditions of cases 8,10) by a factor of $(2.15/4.48 \ 10^{-2})$ equal to 48. The ratio [OIII]/[NII] increases by a factor of 271.

The "line" electron temperature in homogeneous nebulae is larger than the temperature deduced from the corresponding criteria (in Section 11), by a fairly small, and so, negligible amount, except in case of $Z \gg Z_o$ (case 8 in Table 3). Note that only models (of those in Table 3) with low Z (actually $\lesssim 0.5 \ Z_o$) have $T_r(OIII) > T_r(NII)$, otherwise the reverse is true. Again $T_r(OIII) > T_r(NII)$ occurs for low $T_{eff}(\lesssim 30\ 000)$ (Stasinska, 1978). In the last case the effect is attributed to lack of hardening of radiation as one approaches the ionization front, in the first one $(Z \lesssim 0.5 \ Z_o)$ to the fact that at low Z the dominant process for cooling is the recombination of hydrogen which is a decreasing function of the temperature.

A similar behaviour, as with $T_r(OIII)$ and $T_r(NII)$, applies also to planetary nebulae models. Since however in a number of cases we observe $T_r(NII) < T_r(OIII)$, I suspect this indicate a decreasing density outwards in the real nebulae.

Anyhow, apart from cases with $Z \gg Z_o$, the difference between $T_r(NII)$ and $T_r(OIII)$ is small. To neglect it and assume everywhere in a HII region the temperature $T_r(OIII)$, introduces then a not large error in the determination of abundances of low ionization species. It would overestimate such abundances, but only in homogeneous nebulae or in nebulae with an outward increase in density. In the often more realistic cases of an outward decreasing density, the procedure may estimate correctly or possibly underestimate the true abundances of low ionization species.

12.5e Possibility to asses variations of Z from variations in the observed [OIII]/[NII] = R

From Table 3 one sees that $R = \lambda (4959+5007)[OIII]/\lambda(6548+6584)[NII]$ varies little when going from an HII region excited by a single star to a huge complex excited by a stellar association as long as the last is concentrated relative to the gas distribution.

Instead R increases strongly with T_{eff} or with a decrease of Z. Without further information one cannot discriminate between the two possibilities. But if a measurement of T_e is available (and possibly of N_e) one measures Z and so can distinguish between the two effects. On the other hand a decrease of Z in an HII region is likely accompanied by a less blanketed atmosphere of the exciting star. Then T_{eff} is larger and the two effects concur in producing a larger [OIII]/[NII], independently from other causes of larger T_{eff} (as more massive stars).

Thus a Z gradient, in a sequence of HII regions, inferred
assuming constant T_{eff} , is steeper than the one derived considering the gradient of T_{eff} , if the last is present in the indicated direction, i.e. opposite to one of the Z gradient.

This appears to be actually the case for extragalactic HII
regions in spiral galaxies (cf. Section 13.3).

13. CHEMICAL ABUNDANCES IN PHOTOIONIZED NEBULAE AND THEIR GRA-
DIENTS IN GALAXIES

13.1 Planetary Nebulae

Over the about 1280 known galactic planetary nebulae (PN)
(cf. Kohoutek, 1983: IAU Sympos. No. 103) we have information on
chemical abundances in one or more of the following ratios:
He/H, N/H, O/H, N/O for about 100 objects (cf. Kaler, 1979; Kaler
and Hartkopf, 1981), essentially from optical work. These determinations are based on the simplified procedures. Their accuracy
varies somewhat from object to object. Studies in progress,
using the UV spectral range (1150-3200 A) observed with the IUE
satellite, should improve the abundance of N (in the optical seen
only in lines of [NII]) but probably not by large amounts, since
the determination of N/O via the ratio [NII]λ(6584+6548)/[OII]
λ(3727+3729) is relatively accurate.

Analysis of these data to explore the existence of radial
gradients in the Galaxy gives discordant results. According to
Kaler (1983), radial gradients in the Galaxy are not seen in PN,
its possible presence being confused by evolutionary enrichment
processes in He and N and by vertical gradients from the galactic
plane in N/O, He/H, O/H that instead seem to be in little doubt.

On the other hand, analysis of the best observed objects
provides, according to Peimbert and Serrano (1980), a relatively
clear evidence that radial gradients in He/H and N/O are present
in PN. This when objects of the so called types II-III (He/H <
0.125 and N/O < 0.63) are selected. These PN are considered of
intermediate population between type I (He/H > 0.125 or N/O > 1.0
when a He/H determination is lacking) including relatively young
massive objects, and type IV consisting of halo extreme population II objects.

As for the abundance of carbon, it can be obtained from faint
recombination optical lines of C^+ at λ 4267, of C^{2+} at λ 4650 or
from strong collisionally excited lines of CIII] at λ 1909 and of
CIV at λ 1550. The two methods often disagree with the optical

lines giving higher abundances. From studies in points at different radial distance from the central stars (Barker, 1982 in NGC
6720; Barker, 1983 in NGC 7009) it is becoming evident that the
abundances from optical lines are indeed overestimated particularly in regions of high excitation. Results from the UV lines
are therefore more reliable. Studies of this kind have so far
been performed in about 17 objects, mostly with simplified procedure, but even with detailed models or intermediate procedures.

We have selected those belonging to types II-III. They are
12, so that the statistics is comparable with the one of Peimbert
and Serrano (8 objects are in common with them). We have plotted
in Figs. 1 and 2 respectively He/H, log O/H and log N/H, log C/H
versus the galactocentric distance (R_0 is taken to be 9 kpc). The
chemical abundances are all from these recent IUE studies except
He/H taken from the compilation of Kaler (1979), because it is
rarely determined in these studies.

It is evident that the new data scarcely support the existence of radial gradients in He/H or O/H. Indeed the C/H data seem
to indicate, in regions close to the Sun, a strong gradient, opposite to the one indicated by HII regions in our galaxy and in
other spiral galaxies (see later on). Data from objects in the
direction of the anticenter do not conform to this apparent relationship.

Since most of these determinations for heavy elements are
probably accurate to better than a factor of two, some of the
scatter seen in the Figures is probably real, but we feel it is
unsafe to conclude from the present evidence that radial gradients
of chemical composition in our galaxy from planetary nebulae are
proved to exist.

13.2 Galactic HII regions

The most relevant studies of chemical abundances in galactic
HII regions, in order to assess the existence of chemical radial
gradients in our Galaxy, are by Peimbert et al. (1978), Hawley
(1978b), Talent and Dufour (1979). About 20 HII regions have
been investigated with galactocentric distance ranging from 7.9
to 14.1 kpc (R_0 = 10.0 kpc). In most of the 15 HII regions studied by Hawley the electron temperature was not determined from
the observations, but deduced from models. The analysis by Talent
and Dufour uses only objects where T_e could be determined observationally. This work should provide therefore more reliable
abundances. These were derived using the simplified procedures
(Section 12.2). Gradients in O/H and N/H are found to occur in
our galaxy, while there is only marginal evidence for a He/H gradient and little evidence for a S/H gradient. Recently Shaver et

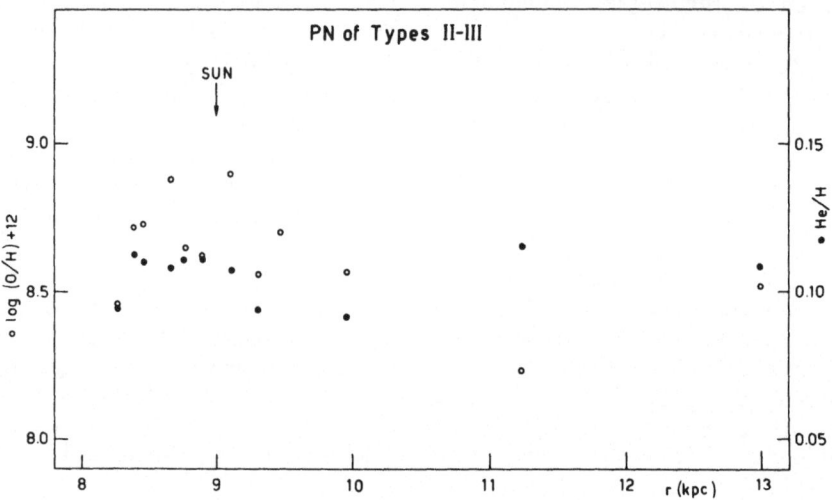

*Fig. 1 – Abundances of helium (dots) and oxygen (open circles)
from planetary nebulae of types II–III (see text) studied so far with the IUE satellite (from various authors)
are plotted versus the galactocentric distance. The sun
is at 9 kpc.*

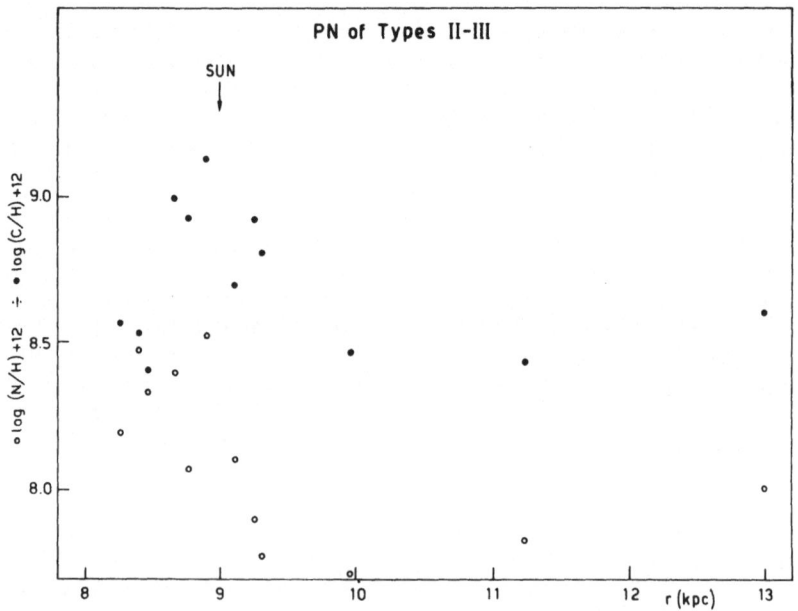

*Fig. 2 – Same as in Fig. 1 for nitrogen (open circles) and carbon
(dots).*

al. (1982) have performed an extended analysis of gradients in galactic HII regions. Electron temperatures were derived with radio recombination lines in 67 HII regions spanning the galactocentric range $3.5 < r_G < 13.7$ kpc. A temperature gradient of

$$T_e = (3150 \pm 110) + (433 \pm 40) \, r_G \quad K$$

was found. These temperatures have been applied to optical spectra of 32 HII regions to investigate the abundance gradients (obtained with the simplified procedures). Gradients for oxygen, nitrogen and argon have been positively found, while for sulfur, neon and He^+/H^+ the evidence is only marginal. Data for the O/H, N/H gradiends from Talent and Dufour and Shaver et al. are compared in Table 4 with values for the well studied spiral galaxy M 101 (see Section 13.3).

13.3 Extragalactic HII regions

We restrict our attention to HII regions in spiral galaxies. The ones more easily observed are typically much larger (~ 100 pc) and of lower density (~ 10 cm^{-3}) than the Orion nebula (~ 5 pc; 300 cm^{-3}). They are excited by a large number of early-type O stars.

First observations of gradients, with the galactocentric distance, in some line intensity ratios, go back to Aller (1942) who noted in M33 the quantity λ (5007+4959)[OIII]/H$_\beta$ increasing outwards from the center of the galaxy, and to Burbidge and Burbidge (1962) who observed λ (6584)[NII]/H$_\beta$ to be larger near the nuclei of spiral galaxies than in the outer galactic disks. Better observations of these and other ratios and a definite interpretation in terms of gradients of chemical abundances came with Searle (1971) and subsequent work by various authors including Shields (1974), Smith (1975), Shields and Tinsley (1976), Shields and Searle (1978), Rayo et al. (1982), Blair et al. (1982).

A list of spirals (13 galaxies) in which the abundances have been relatively well determined from studies of HII regions (and the sources of these studies) is given by Pagel and Edmunds (1981, their Table 4). Among these galaxies, M 101 and M 33 are the best studied, and in the following we will refer essentially to them and more specifically to M101.

To determine unambiguously the chemical composition we have seen (Section 13) one needs to know the electron temperature in the relevant ionization zones. Although this is strictly necessary only when using the "simplified procedures", in practice is needed also when using the detailed model method or the intermediate procedures, in order to reduce the number of little known parameters. In the first accurate observations (Searle, 1971), and often in subsequent work on various HII regions of different galaxies, the electron temperature was not directly determined since neither of the two usually most relevant faint optical lines $\lambda 4363$[OIII], $\lambda 5755$ [NII] were measured. Actually the observations show clear trends in the intensity ratios $R_{OIII} = \lambda(4959+5007)$[OIII]/$H_\beta$, $R_{NII} = \lambda(6548+6584)$[NII]/$H_\alpha$ and $R_3 = (3727+29)$[OII]/$\lambda(6548+84)$[NII] with the galactocentric distance of the considered HII regions.

R_{OIII} increases outwards by 1.5-2 order of magnitudes (in the mentioned M101 and M33), R_{NII} decreases outwards by one order of magnitude and R_3 increases outwards by about 2 orders of magnitudes. The spectra of the HII regions were seen not to depend in any systematic way on the appearance, luminosity or surface brightness of the regions. The behaviour of R_{OIII} implies that inner HII regions have lower excitation. Excluding systematic variations in the geometry or in mass distribution of the gas in the different HII regions, this could be due to one or more of the following circumstances:

a) lower temperature of the ionizing sources of inner HII regions, so that a smaller volume of the nebula is ionized from O^+ to O^{++},
b) higher dust content in the inner HII regions, able to absorb radiation beyond the Lyman limit,
c) higher abundance of O/H in the inner HII regions, O^{2+} being the most important cooler in the He^+ zone.

The behaviour of R_{NII} would be consistent, in each of the three cases, with an outward increase of [NIII]/[NII], similar to the one in fact observed for $\lambda(5007)$[OIII]/$\lambda(3727)$[OII] (this varies by an order of magnitude). The behaviour of R_3 suggests an outward increase of O/N and/or of T_e . In all cases a, b, c, the temperature in the inner regions should be lower than that in the outer arms of the galaxies. Under hypothesis a and b the HeI recombination lines should increase outwards, because of the reduction, in the inner regions, of photons able to ionize He° relative to the H Ly_c photons. Instead hypothesis c does not imply variations in the recombination lines HeI/HI, that, if present, should be attributed, in this case, to a gradient of He/H.

Searle (1971) using a simple HII region model, supported the explanation c and this was confirmed as the most important cause of the excitation gradient in all subsequent work.

In particular Smith (1975) has been able to measure also the $\lambda 4363$[OIII], $\lambda 5876$ HeI, $\lambda 6716+31$ [SII] and $\lambda 3869$ [NeIII] lines. He finds that the electron temperature regularly increases from 8800 K (in NGC 5461 at 5'.2 of angular distance from the cen-

ter of M101) to 14 000 K (in NGC 5471 at 12'.2). Thus the abundances were determined unhambiguously for these regions. Moreover
he observes that λ 5876 HeI/H$_\beta$ is practically constant, and that
also the Balmer decrement does not show systematic radial variations, supporting the view that a and b cannot be the main causes
of the observed excitation gradient.

Actually an increase of T_u , the temperature of the hottest
stars present, of $\Delta \log T_u$ = 0.02 - 0.13 from the intermediate to
outermost spiral arms of M101 was found necessary to explain the
observed outward increase of the equivalent width of H$_\beta$ (by a
factor of $\sim$ 3) (Shields and Tinsley, 1976). Such a gradient in
T_u coupled with model atmospheres having the same metal abundance as the nebular gas (and so accounting for metal absorption
edges in the radiation of ionizing stars) was found by Shields
and Searle (1978) (using detailed models) to reproduce fairly well
accurate observations of three selected HII regions in M101.
(Without such improvements the [OII] lines were observed to be too
strong in the low excitation inner region).

The recent work by Rayo et al. (1982) follows the same line
of observing very accurately 5 HII regions in M101. They use the
simplified method to deduce chemical abundances, and in particular do confirm the outward increasing gradient in T_e . They obtain $\Delta T_e / \Delta R$ = 295±30 K kpc^{-1}, in agreement with values derived
with radio telescopes for galactic HII regions in the 4-14 kpc
range by various authors (e.g. Churchwell et al., 1978, Wilson et
al. 1979) and moderately smaller than the most recent value from
radio recombination lines amounting to 433±40 K kpc^{-1} (Shaver et al.,
1982).
     In conclusion negative gradients with galactocentric distan-
ce in spiral galaxies from studies of HII regions appear well
established for O/H and N/H. A moderate gradient of He/H is possibly indicated (Smith, 1975; Shields and Searle, 1978; Rayo et
al., 1982) but the existence of gradients of Ne/O, S/O, Ar/O in
external spirals (claimed by Rayo et al.) is, in our opinion,
still to be proved.

We have reported in Table 4 what we believe are the best values for the N and O abundances in M101. Also in Table 4, data
for our galaxy by Talent and Dufour (1979) and by Shaver et al.
(1982) are reported. All the data have been reevaluated from the
original abundances (some misprints have been occasionally noted
in the quoted values of the gradients) and the correlation coefficients of the least square fits are shown in parenthesis.

In M101 it appears that the (absolute) gradients of O/H and
particularly of N/H are smaller in the recent work of Rayo et al.
(1982) than from previous works.

TABLE 4. RADIAL GRADIENTS IN SPIRAL GALAXIES

Galaxy	$\dfrac{d \log O/H}{d \log \rho}$	$\dfrac{d \log O/H}{d\rho}$	$\left\langle \dfrac{\log N/O}{\log O/H} \right\rangle$	$\dfrac{d \log N/H}{d \log \rho}$	$\dfrac{d \log N/H}{d\rho}$	$\dfrac{d \log O/H}{dR(kpc)}$	$\dfrac{d \log N/H}{dR(kpc)}$
M 101 Ref.							
(1) 6 regions ρ (5'.2–12'.2)	− 1.82 (0.88)	− 0.10 (0.95)	0.39±0.05	–	–	− 0.046 (0.95)	–
(2) 3 regions ρ (3'.2–12'.2)	− 1.60 (0.94)	− 0.095 (0.85)	0.34±0.07	− 2.99 (0.98)	− 0.18 (0.92)	− 0.044 (0.85)	− 0.083 (0.92)
(3) 3 regions ρ (3'.4–12'.2) = (7.3–26.2)kpc	− 1.12 (0.93)	− 0.072 (0.99)	0.40±0.01	− 1.73 (0.97)	− 0.11 (1.00)	− 0.033 (0.99)	− 0.051 (1.00)
Our Galaxy							
(4) 11 regions R (7.9–14.1)kpc	–	–	0.30±0.05	–	–	− 0.059 (0.36)	− 0.063 (0.38)
(5) 19 regions R (7.1–13.7)kpc	–	–	0.35±0.04	–	–	− 0.064 (0.46)	− 0.079 (0.65)

Ref. (1) Smith, 1975; (2) Shields, Searle (1978); (3) Rayo et al. (1982); (4) Talent, Dufour (1979); (5) Shaver et al. (1982).

In parenthesis are the correlation coefficients r^2.

The absolute gradients of O/H, and also of N/H, if results
by Rayo et al. are preferred, as we do, result larger in our
galaxy than in M101. A modest gradient of N/O increasing inwards
is also indicated in both galaxies, suggesting that a little fraction of the observed N might be produced by secondary undersynthe-
sis processes.

These quantitative comparisons are however still to be regar-
ded with caution because the regions well studied in M101 are quite a few and the study of galactic HII regions shows that local
scatter may be important. On the other hand data in our galaxy
cover a smaller galactocentric distance of 6 kpc compared with 19
kpc in M101.

REFERENCES

Aller, L.H.: 1942, Astrophys. J. 95, 52.
Aller, L.H. and Czyzak, S.J.: 1982, U.C.L.A. Astron. Astrophys.
    Preprint No. 140.
Aller, L.H. and Liller, W.: 1959, Astrophys. J. 130, 45.
Alloin, D., Collin-Souffrin, S., Joly, M. and Vigroux, L.: 1979,
    Astron. Astrophys. 78, 200.
Balick, B. and Sneden, C.: 1976, Astrophys. J. 208, 336.
Barker, T.: 1980, Astrophys. J. 240, 99.
Barker, T.: 1982, Astrophys. J. 253, 167.
Barker, T.: 1983, Astrophys. J., in press.
Beck, S.C., Lacy, J.H., Townes, C.H., Aller, L.H., Geballe, T.R.,
    Baals, F.: 1981, Astrophys. J. 249, 592.
Benvenuti, P. and Perinotto, M.: 1981, Astron. Astrophys. 95, 127.
Blair, W.P. and Kirshner, R.P.: 1982, Astrophys. J. 254, 50.
Burbidge, E.M. and Burbidge, G.R.: 1962, Astrophys. J. 135, 694.
Churchwell, E., Smith, L.F., Mathis, J. and Mezger, P.G.: 1978,
    Astron. Astrophys. 70, 719.
Dinerstein, H.L.: 1983, Proceed. IAU Symp. No. 103, "Planetary
    Nebulae", p. 79.
French, H.B.: 1981, Astrophys. J. 246, 434.
French, H.B. and Grandi, S.A.: 1981, Astrophys. J. 244, 493.
Grandi, S.A. and Hawley, S.A.: 1978, Publ. A.S.P. 90, 125.
Harrington, J.P., Seaton, M.J., Adams, S. and Lutz, J.H.: 1982,
    Monthly Not. Roy. Astron. Soc. 199, 517.
Harrington, J.P. and Feibelman, W.A.: 1982, Astrophys. J., in
    press.
Hawley, S.A.: 1978a, Publ. A.S.P. 90, 370.
Hawley, S.A.: 1978b, Astrophys. J. 244, 417.
Hawley, S.A. and Miller, J.S.: 1977, Astrophys. J. 212, 94.
Hawley, S.A. and Miller, J.S.: 1978, Publ. A.S.P. 90, 39.
Kaler, J.B.: 1976, Astrophys. J. Suppl. 31, 517.
Kaler, J.B.: 1979, Astrophys. J. 228, 163.
Kaler, J.B.: 1983, Proceed. of IAU Symp. No. 103, "Planetary Ne-

bulae", p. 245.

Kaler, J.B., Aller, L.H. and Bowen, I.S.: 1965, Astrophys. J. 141, 912.

Kaler, J.B. and Kartkopf, W.I.: 1981, Astrophys. J. 249, 602.

Kohoutek, L.: 1983, Proceed. of IAU Symp. No. 103, "Planetary Nebulae", p. 17.

Koppen, J. and Wehrse, R.: 1982, Proceed. of Third European IUE Conference, Madrid, May 1982, in press.

Lucy, J.H., Beck, S.C. and Geballe, T.R.: 1982, Astrophys. J. 255, 510.

Mathis, J.S.: 1970, Astrophys. J. 159, 263.

Mathis, J.S.: 1978, "Planetary Nebulae", IAU Symp. No. 76, Y. Terzian, ed. (Dordrecht: Reidel), p. 281.

Mathis, J.S., Perinotto, M., Patriarchi, P. and Schiffer, F.H.: 1981, Astrophys. J. 249, 99.

Mihalas, D.: 1972, "Non-LTE Model Atmospheres for B and O stars", NCAR-TN/STR-76.

Pagel, B.E.J., Edmunds, M.G., Blackwell, D.E., Chun, M.S. and Smith, G.: 1979, Monthly Not. Roy. Astron. Soc. 189, 95.

Pagel, B.E.J. and Edmunds, M.G.: 1981, Ann. Rev. Astron. Astrophys. 19, 77.

Peimbert, M. and Costero, R.: 1969, Bol. Obs. Tonantzintla y Tacubaya 5, 3.

Peimbert, M. and Torres-Peimbert, S.: 1971, Astrophys. J. 168, 413.

Peimbert, M. and Torres-Peimbert, S.: 1977, Monthly Not. Roy. Astron. Soc. 179, 217.

Peimbert, M., Torres-Peimbert, S. and Rayo, J.F.: 1978, Astrophys. J. 220, 516.

Peimbert, M. and Serrano, A.: 1980, Rev. Mex. Astron. Astrof. 5, 9.

Pequignot, D., Aldrovandi, S.M.V. and Stasinska, G.: 1978, Astron. Astrophys. 63, 313.

Perinotto, M.: 1971, Astron. Astrophys. 14, 78.

Perinotto, M.: 1977, Astron. Astrophys. 61, 247.

Perinotto, M. and Patriarchi, P.: 1980, Astrophys. J. 235, L13.

Rayo, J.F., Peimbert, M. and Torres-Peimbert, S.: 1982, Astrophys. J. 255, 1.

Sarazin, C.L.: 1977, Astrophys. J. 211, 722.

Savage, B.D. and Mathis, J.S.: 1979, Ann. Rev. Astron. Astrophys. 17, 73.

Searle, L.: 1971, Astrophys. J. 168, 327.

Shaver, P.A., McGee, R.X., Newton, L.M., Danks, A.C. and Pottasch, S.R.: 1982, preprint.

Shields, G.A.: 1974, Astrophys. J. 193, 335.

Shields, G.A. and Tinsley, B.M.: 1976, Astrophys. J. 203, 66.

Shields, G.A. and Searle, L.: 1978, Astrophys. J. 222, 821.

Simpson, J.P.: 1975, Astron. Astrophys. 39, 43.

Smith, H.E.: 1975, Astrophys. J. 199, 591.

Stasinska, G.: 1978, Astron. Astrophys. 66, 257.

Stasinska, G.: 1980, Astron. Astrophys. 84, 320.

Stasinska, G., Alloin, D., Collin-Souffrin, S. and Joly, M.: 1981,
    Astron. Astrophys. 93, 362.
Talent, D.L. and Dufour, R.J.: 1979, Astrophys. J. 233, 888.
Wilson, T.L., Pauls, T.A. and Ziurys, L.M.: 1979, Astron. Astro-
    phys. (Letters) 77, L3.

REVIEW OF ATOMIC PHYSICS AS APPLIED TO THE DIFFUSE INTERSTELLAR MEDIUM

D.R. Flower

Physics Department, The University, Durham DH1 3LE, U.K.

ABSTRACT

Atomic physics processes relevant to the ionisation and thermal balance of the diffuse interstellar medium are reviewed. Particular attention is paid to charge transfer reactions, dielectronic recombination, and to photoionisation and radiative recombination. Collisional excitation of fine structure transitions by electrons and hydrogen atoms is also considered in relation to the cooling of the interstellar gas.

1. INTRODUCTION

The field of atomic physics may sometimes appear as a maze to the uninitiated astronomer. The ritual of initiation consists of learning that it is not only a maze but a minefield! It is my purpose, in this review, to provide some guidance through those parts of the maze which must be explored in order to study the interstellar medium (ISM). I shall concentrate on those physical processes which establish the ionisation and thermal equilibria of the diffuse ISM.

J. Audouze et al. (eds.), Diffuse Matter in Galaxies, 239–257.

2. CHARGE TRANSFER REACTIONS

The recognition of the importance of charge transfer (CT) in the interstellar medium was probably initiated by Field and Steigman (1) who considered the reaction

$$O^+ + H^0 \overset{\rightarrow}{\leftarrow} O^0 + H^+ + \Delta E \tag{2.1}$$

Earlier work of Chamberlain (2) had shown that (2.1) was likely to be the dominant reaction in establishing the ionisation equilibrium of O^0 in the Cassiopeia radio source. Williams (3) subsequently demonstrated the importance of this same reaction in the planetary nebula IC 418.

The rate coefficient for reaction (2.1) may be expected to be large ($\approx 10^{-9}$ cm^3 s^{-1}) at thermal energies because of the near coincidence of the ionisation potentials of O^0 and H^0 ($\Delta E = 0.020$ eV $\equiv 232$ K). Most CT reactions do not satisfy this near resonance condition and the evaluation of the corresponding rate coefficients poses subtle problems.

2.1. The formula of Landau and Zener

Bates and Moiseiwitsch (4) made the first quantitative evaluations of cross-sections for CT between positive ions and atomic hydrogen. They employed a formula derived independently by Landau and Zener and by Stueckelberg (the derivations are given by Mott and Massey (5)).

Consider the reaction

$$X^{m+} + H^0 \overset{\rightarrow}{\leftarrow} X^{(m-1)+} + H^+ + \Delta E, \tag{2.2}$$

where X^{m+} denotes the element X, m-times ionised. In the $X^{m+} + H^0$ channel, the long range interaction is attractive owing to the polarisation of the hydrogen atom by the ion,

$$V_{pol}(r) \sim -\tfrac{1}{2} \alpha m^2/r^4, \tag{2.3}$$

where r is the internuclear distance, α is the polarisability of H^0, and atomic units are employed. If m > 2, the Coulomb repulsion between the residual ion and the proton determines the long range interaction in the $X^{(m-1)+} + H^+$ channel; in atomic units,

$$V_{Coul}(r) \sim (m-1)/r. \tag{2.4}$$

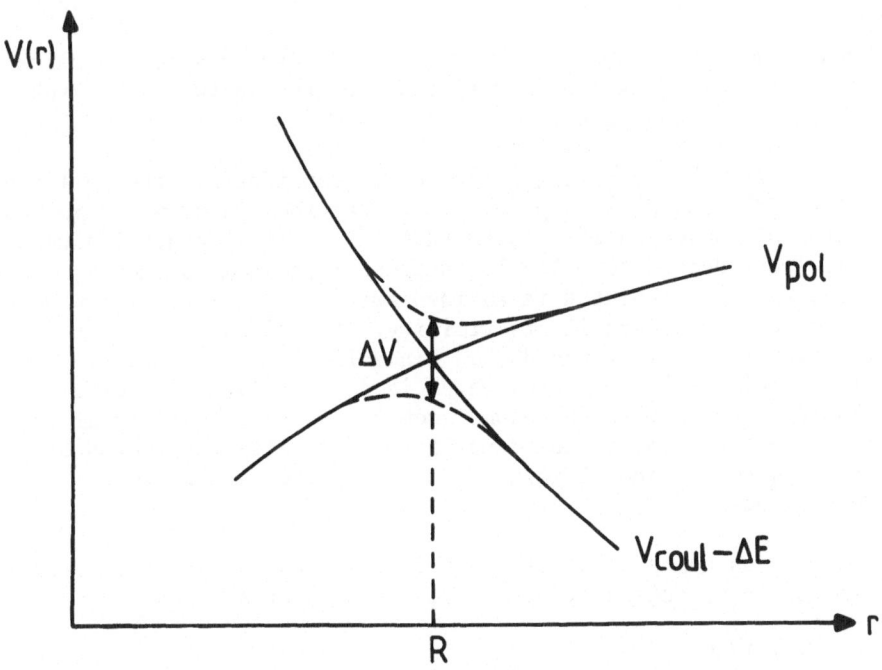

Figure 1. Long range potential energy curves in the $X^{m+} + H^0$ channel (V_{pol}) and in the $X^{(m-1)+} + H^+$ channel (V_{Coul}); ΔE is the energy defect of the charge transfer reaction (cf. Equ. 2.2). The crossing of the continuous ("diabatic") potential curves at the point R is avoided owing to the interaction between the corresponding molecular states, which are assumed to have the same symmetry. The dashed ("adiabatic") curves are then obtained, separated by an energy $\Delta V(R)$ at the avoided crossing.

In Fig. 1, we sketch the corresponding, long range potential energy curves, which cross at the point R, given by

$$V_{Coul}(R) - \Delta E = V_{pol}(R),\tag{2.5}$$

where ΔE is the energy defect in reaction (2.2). If the crossing occurs at sufficiently long range, $V_{pol}(R) \ll \Delta E$ and hence

$$V_{Coul}(R) = (m-1)/R \approx \Delta E \qquad (2.6)$$

From (2.6), we obtain a simple but useful expression for the crossing distance,

$$R \approx (m-1)/\Delta E. \qquad (2.7)$$

When atomic units are employed, distances are measured in units of the Bohr radius ($a_0 \approx 0.5292$ A) and energies in twice the Rydberg constant ($2R_\infty \approx 27.2$ eV).

At thermal energies, which we are considering, the collision between X^{m+} and H^0 is appropriately described in terms of molecular states of the composite system $(XH)^{m+}$. If the potential curves in Fig. 1 correspond to molecular states of the same symmetry, the crossing at the point R is avoided owing to the interaction between the molecular states in its vicinity. The adiabatic potential energy curves follow the dashed lines in Fig. 1 and are separated at the point R by an energy which is denoted $\Delta V(R)$. Charge transfer occurs through a jump from one adiabatic potential curve to the other. The Landau-Zener formula enables the probability of such a jump, in the vicinity of an avoided crossing, to be determined.

Let us compare the relevant time scales. The characteristic time scale of internal motion, τ, may be expressed as

$$\tau \approx 1/\Delta V(R), \qquad (2.8)$$

where the atomic unit of time is 2.42×10^{-17}s. The effective collision or interaction time may be estimated as the time required for the particles to separate by a distance ΔR from the point of avoided crossing, such that the energy gap between the adiabatic potential energy curves is approximately doubled. Performing a Taylor series expansion around $r = R$, we obtain

$$\Delta V(R) \approx |V'_{Coul}(R) - V'_{pol}(R)| \, \Delta R \qquad (2.9)$$

where the primes denote differentiation with respect to r. Hence,

$$\Delta R \approx \Delta V(R) / |V'_{Coul}(R) - V'_{pol}(R)| \qquad (2.10)$$

and the collision time is

$$T \approx \Delta R/v_\ell(R)$$
$$\approx \Delta V(R) / \left(v_\ell(R) |V'_{Coul}(R) - V'_{pol}(R)| \right) \qquad (2.11)$$

where $v_\ell(R)$ is the radial component of velocity at $r = R$ in a state of relative orbital angular momentum, ℓ. Comparing τ and T, we have

$$T/\tau \approx \Delta V^2 / \left(v_\ell \left| V'_{Coul} - V'_{pol} \right| \right), \tag{2.12}$$

where all quantities are to be evaluated at $r = R$. If $T/\tau \gg 1$ (slow collision), the particles follow the adiabatic (dashed) potential curves in Fig. 1. If $T/\tau \ll 1$ (fast collision), the diabatic (continuous) curves are followed. The probability of CT is small in both of these limits.

The Landau-Zener formula asserts that the probability of a jump from one adiabatic potential energy curve to the other in a single pass through the avoided crossing is

$$P_\ell = e^{-w}, \tag{2.13}$$

where

$$w = \frac{\pi}{2} \frac{T}{\tau}$$
$$= \pi \, \Delta V^2 / \left(2 v_\ell \left| V'_{Coul} - V'_{pol} \right| \right) \tag{2.14}$$

and all dimensional quantities are in atomic units. The overall probability of CT in a given collision is

$$P_\ell = 2 P_\ell (1 - P_\ell) \tag{2.15}$$

Referring to Equs. (2.14) and (2.15), we see that $P_\ell \approx 0$ when $T/\tau \gg 1$ or $T/\tau \ll 1$, as stated above.

The CT cross-section is then given by

$$Q = \frac{\pi}{k_i^2} \sum_{\ell=0}^{L} (2\ell+1) \, P_\ell, \tag{2.16}$$

where k_i is the wave vector in the incident channel, i, and L is determined by the condition that $v_L(R) = 0$.

The sum in (2.16) may be transformed into an integral,

$$Q = 2\pi \int P(b) \, b \, db, \tag{2.17}$$

where b is the impact parameter. Transformation to a new integration variable,

$$x = \left(1 - b^2 / \left(R^2 (1+\lambda) \right) \right)^{-\frac{1}{2}} \tag{2.18}$$

yields

$$Q = 2\pi \; R^2(1+\lambda) \int_1^\infty P(x) \; x^{-3} \; dx, \tag{2.19}$$

where

$$\lambda = \left(V_i(\infty) - V_i(R)\right)/E_i, \tag{2.20}$$

E_i being the incident energy, and

$$P(x) = 2 \; e^{-\eta x}(1 - e^{-\eta x}) \tag{2.21}$$

with

$$\eta = \pi \; M \; \Delta V^2 / \left(2k_i \; (1+\lambda)^{\frac{1}{2}} \; |V'_{Coul} - V'_{pol}|\right) \tag{2.22}$$

M being the reduced mass of the system (in atomic units).

Butler and Dalgarno (6) have discussed the evaluation of CT cross-sections from the Landau-Zener formula. They claim that, for simple transfer of the 1s electron of H^0 to an orbital of the heavy ion, X^{m+}, the energy gap $\Delta V(R)$ may be represented by

$$\Delta V(R) = R^2 \; e^{-R}, \tag{2.23}$$

where ΔV and R are expressed in atomic units. Thus, if R is known (e.g. from Equ. (2.7)), ΔV may be determined from (2.23). This equation also tells us that, if R is small, ΔV is large and the collision will tend to proceed along the adiabatic potential curves. Conversely, if R is large, ΔV is small and a diabatic collision occurs. As noted above, the CT cross-section is small in both these limits.

In the asymptotic region,

$$|V'_{Coul}(R) - V'_{pol}(R)| = (m-1) \; (1+\mu)/R^2 \tag{2.24}$$

where

$$\mu = R^2 \; V'_{pol}(R)/(m-1) \tag{2.25}$$

is a correction term arising from the polarisation potential (2.3). Equs. (2.23-2.25) provide the additional information necessary to evaluate the Landau-Zener CT cross-section, Q.

Butler and Dalgarno (6) compare rate coefficients (at $T = 10^4K$) evaluated from the Landau-Zener formula with the results of quantum mechanical calculations. They consider the formula to be remarkably accurate when the condition $E_i(R) >> \Delta V(R)$ is satisfied.

2.2. Orbiting approximation

Bates (7) has shown that $E_i(R) >> \Delta V(R)$ is a necessary condition for the Landau-Zener approximation to be valid. It follows that the approximation progressively breaks down as the collision energy, E_i, decreases. The failure of the method is illustrated in Fig. 2 for the CT reaction.

$$Si^{2+}(3s^2\ {}^1S) + H^0 \rightarrow Si^+(3s^2\ 3p\ {}^2P^0) + H^+ + 2.74\ eV \qquad (2.26)$$

(McCarroll and Valiron (8)). The Landau-Zener formula grossly underestimates the CT cross-section at low energies. The reason for the failure is the neglect of classical trajectory effects, particularly orbiting (McCarroll and Valiron (8); Gargaud et al. (9)).

Consider the classical energy conservation equation,

$$E_i = \tfrac{1}{2} M v_\ell^2(r) + E_i\ b^2/r^2 - 2\alpha/r^4, \qquad (2.27)$$

where we take the incident channel, i, to be $Si^{2+} + H^0$. At the distance of closest approach, r_c, $v_\ell(r_c) = 0$ and hence

$$E_i = E_i\ b^2/r_c^2 - 2\alpha/r_c^4. \qquad (2.28)$$

Taking the derivative of (2.28) with respect to r_c, we obtain

$$2E_i\ b^2/r_c^3 = 8\alpha/r_c^5, \qquad (2.29)$$

which is the classical condition for orbiting to occur, when the centrifugal and induction forces balance. From (2.28) and (2.29), we obtain the corresponding expression for the impact parameter,

$$b = (8\alpha/E_i)^{\tfrac{1}{4}}; \qquad (2.30)$$

for this value of b, the Si^{2+} ion and the H^0 atom will remain together for a relatively long time, enhancing the probability of charge transfer. Taking $P = \tfrac{1}{2}$ for such an encounter, we obtain $P = \tfrac{1}{2}$ (its maximum value). The CT cross-section may then be estimated as

$$Q \approx \tfrac{1}{2}\ \pi \left(\frac{8\alpha}{E_i}\right)^{\tfrac{1}{2}}. \qquad (2.31)$$

In Table 1, we compare values of Q derived from (2.31) for $E_i \lesssim \Delta V = 0.07\ eV \equiv 812\ K$ with the results of quantal calculations by McCarroll and Valiron (8). The agreement is, perhaps surprisingly, good.

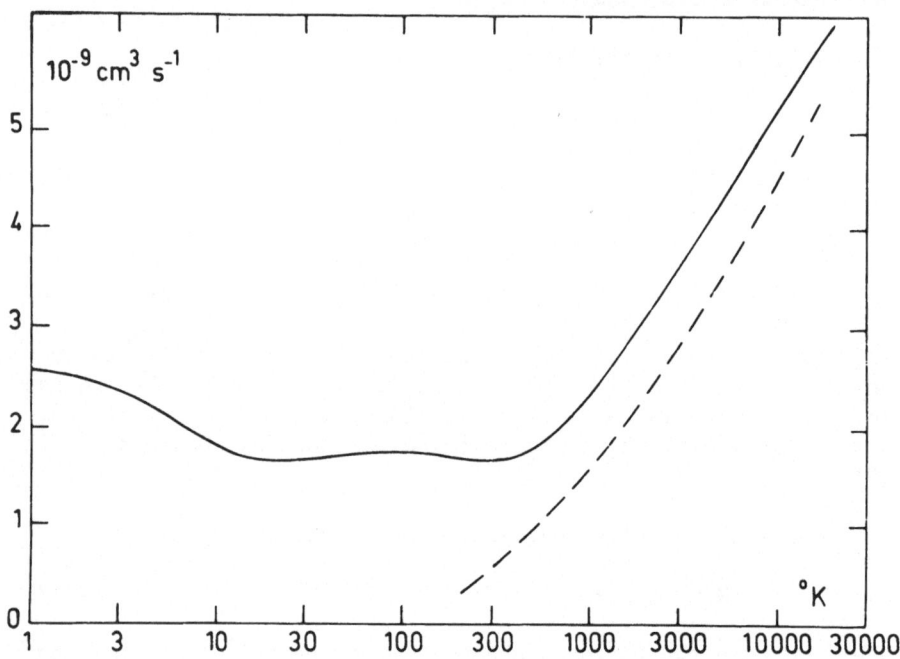

Figure 2. Computed rates of the reaction $Si^{2+} + H^0 \rightarrow Si^+ + H^+ +$ 2.74 eV: continuous curve, quantal calculations of McCarroll and Valiron (8); dashed curve, Landau–Zener calculation. For this reaction, the separation of the potential energy curves at the avoided crossing, $\Delta V(R) = 812$ K.

The rate coefficient corresponding to (2.31) may be derived from

$$q(T) = \left(\frac{8kT}{\pi M}\right)^{\frac{1}{2}} \int_0^\infty y \, Q(y) \, e^{-y} \, dy, \qquad (2.32)$$

where

$$y = E_i/(kT). \qquad (2.33)$$

Performing the integral, we obtain

$$q = 2\pi(\alpha/M)^{\frac{1}{2}}, \qquad (2.34)$$

which is the <u>Langevin</u> rate (independent of T).

E_i/k (K)	Q (a_0^2)	
	(1)	(2)
1	6980	5296
3	3730	3058
10	1570	1675
30	826	967
100	472	530
300	259	306
1000	197	167

Table 1. Cross sections, Q, for the charge transfer reaction
$Si^{2+} + H^0 \rightarrow Si^+ + H^+$ as a function of incident energy, E_i:
(1) McCarroll and Valiron (8); (2) estimation from Equ. (2.31).

2.3. Quantum mechanical calculations

Whilst simple (Landau-Zener and orbiting approximations) can
yield reasonable estimates of the rates of CT reactions of type
(2.2), results so obtained should be interpreted as an indication
rather than a proof of the possible importance of a CT process.
Quantum mechanical calculations of the relevant potential energy
curves and CT cross-sections are the established means of
determining the corresponding rate coefficients. In Table 2, we
summarise the results of recent calculations for a number of
important CT processes.

3. DIELECTRONIC RECOMBINATION

Dielectronic recombination may be visualised as a two-step
process, the first part of which is

$$X^{m+} + e \rightarrow \left(X^{(m-1)+}\right)^*, \qquad (3.1)$$

where * denotes an autoionising state of the recombined ion,
$X^{(m-1)+}$; autoionisation is the inverse of the recombination
process (3.1). Applying the principle of detailed balance, we
obtain

$$\gamma = \frac{\omega_*}{2\omega_+} \left(\frac{h^2}{2\pi mkT}\right)^{\frac{3}{2}} e^{-E_*/(kT)} A_a \qquad (3.2)$$

Rate coefficient (10^{-9} cm^3 s^{-1})

Ion	T=10	30	50	10^2	300	10^3	3000	5000	10^4	50 000 K
C^{3+} $\lbrace$		1.5		1.6	1.6	1.6	1.6		1.6 (12)	
								3.09	3.58	5.46 (10)
C^{4+}	3.39	3.12		2.71	2.38	2.25	2.19		2.13	3.22 (9)
N^{2+}								0.78	0.86	1.11 (10)
N^{3+} $\lbrace$	0.35	0.30		0.25	0.25	0.43		1.54	2.93	9.47 (10)
O^{+}	0.34		0.37	0.41		0.75 (13)	1.12	1.82	3.41	11.23 (9)
O^{2+} $\lbrace$ H^0								0.60	0.77	1.62 (10)
He^0								0.10	0.20	0.89 (10)
O^{3+}								6.34	8.63	17.6 (10)
Ne^{3+}								4.00	5.68	13.0 (10)
Si^{2+}	1.98		1.75	1.72		2.50		4.34	5.28	7.70 (11)

Table 2. Calculated rate coefficients for charge transfer reactions with H^0 (unless otherwise indicated). Numbers in parentheses are references.

for the rate coefficient of process (3.1), where ω_+ and ω_* are the statistical weights of the initial and final atomic states, respectively. The autoionising state is located at an energy E_* above the ground state of X^{m+}, and A_a is the autoionisation probability.

The process of dielectronic recombination is completed by radiative stabilisation,

$$\left(X^{(m-1)+}\right)^* \rightarrow X^{(m-1)+} + h\nu, \tag{3.3}$$

and the dielectronic recombination coefficient is

$$\alpha_d = \gamma \, A_r / (A_r + A_a) \tag{3.4}$$

where A_r is the radiative transition probability.

At high (e.g. coronal) temperatures, it is necessary to consider the full Rydberg series of autoionising levels contributing to the total rate of dielectronic recombination. Radiative stabilisation then occurs principally through the process of core relaxation, and the general formula derived by Burgess (14) is applicable. However, at lower temperatures (more typical of, for example, planetary nebulae), the general formula greatly underestimates the dielectronic recombination rate: radiative stabilisation through transitions of the <u>captured</u> electron must then be considered.

Storey (15) has calculated coefficients for dielectronic recombination to form C^+, C^{2+}, N^{2+}, N^{3+}, and O^{4+} in the temperature range $0.7 \times 10^4 \lesssim T \lesssim 1.5 \times 10^4 K$. Further calculations for other ions are in hand. Storey notes that the expressions for dielectronic recombination rate coefficients given by Aldrovandi and Pequignot (16) are based upon the general formula of Burgess (14) and grossly underestimate the true rates of dielectronic recombination at these temperatures.

It may be recalled that the relative importance of charge transfer and dielectronic recombination is determined by the ratio

$$\frac{n(X^{m+})\ n(H^0)\ q(T)}{n(X^{m+})\ n_e\ \alpha_d(T)} = \frac{n(H^0)}{n_e}\ \frac{q(T)}{\alpha_d(T)} \tag{3.5}$$

and depends on the degree of ionisation of the medium as well as the relative magnitude of the recombination coefficients.

4. PHOTOIONISATION AND RADIATIVE RECOMBINATION

We consider the inverse processes,

$$X^{m+} + h\nu \rightleftarrows X^{(m+1)+} + e. \tag{4.1}$$

Burgess (17) has evaluated photoionisation cross-sections and radiative recombination coefficients for hydrogen, and a FORTRAN programme which evaluates hydrogenic recombination coefficients has been published by Flower and Seaton (18). The hydrogenic approximation is appropriate for recombination not only with fully-stripped nuclei but also to excited states of ions possessing bound electrons, providing that the appropriate value of the screened nuclear charge is employed. Harrington et al. (19) have given in their Table 13 a useful list of references to photo-ionisation cross-sections of ions of the more abundant elements: He, C, N, O, Ne, Mg, Si, and S.

Recent calculations, based upon the close coupling method, take account not only of direct photoionisation but also of the autoionisation process. As an example, consider the photoionisation of N^0: this may proceed directly,

$$N^0(2s^2 \, 2p^3 \, {}^4S^0) + h\nu \rightarrow N^+(2s^2 \, 2p^2 \, {}^3P) + k \left\{ \begin{matrix} s \\ d \end{matrix} \right. \tag{4.2}$$

where k denotes the wave number of the ejected electron and s(d) denotes its angular momentum state; alternatively, ionisation may occur via autoionising levels situated in the first ionisation continuum ($h\nu \gtrsim 14.5$ eV), e.g.

$$N^0(2s^2 \, 2p^3 \, {}^4S^0) + h\nu \rightarrow N^0(2s \, 2p^3({}^5S^0) \, np \, {}^4P)$$

$$\rightarrow N^+(2s^2 \, 2p^2 \, {}^3P) + k \left\{ \begin{matrix} s \\ d \end{matrix} \right. \tag{4.3}$$

The process of autoionisation gives rise to series of resonances in the photoionisation cross-section (Le Dourneuf et al. (20,21)). The contribution of the resonances to the total rate of photo-ionisation depends on their positions and widths and on the spectral distribution of the ionising radiation.

Close coupling calculations of cross-sections for photo-ionisation of C^0 (Burke and Taylor (22)), O^0 (Taylor and Burke (23)), and Ne^0 (Luke (24)) are available. Results of similar calculations for neon ions (Ne^+, Ne^{2+} and Ne^{3+}) have been published by Pradhan (25,26) and for C^{2+} and C^{3+} by Drew and Storey (27).

As noted above, radiative recombination is the inverse of photoionisation, and the recombination coefficient may be derived from the photoionisation cross-section a_ν through the principle of detailed balance,

$$\alpha_i(T) = \frac{1}{c^2} \left(\frac{2}{\pi}\right)^{\frac{1}{2}} (mk\,T)^{-\frac{3}{2}} \frac{\omega_i}{\omega_+} e^{I_i/(kT)}$$

$$x \int_{I_i}^{\infty} (h\nu)^2 \, a_\nu \, e^{-h\nu/(kT)} \, d(h\nu), \tag{4.4}$$

where ω_i is the statistical weight of the recombined ion, ω_+ the statistical weight of the recombining ion, and I_i is the ionisation energy of the state i.

Consider Equ. (4.3): the initial and autoionising states are seen to be connected by a radiative transition, whereas the auto-ionising and final states are connected by a radiationless (collisional) transition. In the notation introduced in Section 3 above, this corresponds to the assumption that the autoionisation probability, $A_a \gg A_r$, the radiative transition probability, in which case the dielectronic recombination coefficient may be written

$$\alpha_d = \frac{\omega_*}{2\omega_+} \left(\frac{h^2}{2\pi mkT}\right)^{\frac{3}{2}} e^{-E_*/(kT)} A_r, \tag{4.5}$$

i.e. dielectronic recombination assimilates to a radiative process (the value of the rate coefficient is determined by the radiative transition probability, A_r). It follows that, if the resonances arising from autoionising levels in the ionisation continuum are included in the integral over a_ν, no further correction to (4.4) is necessary to account for dielectronic recombination. The condition $A_a \gg A_r$ is often assumed (e.g. Storey (15)) but may always be valid, in which case (4.4) does not correctly incorporate dielectronic recombination.

Expressions for radiative recombination coefficients have been derived by Tarter (28,29), Aldrovandi and Pequignot (16), and Gould (30). In none of these compilations are resonances in the photo-ionisation cross-sections taken into account and further corrections for dielectronic recombination are necessary.

5. FINE STRUCTURE EXCITATION CROSS-SECTIONS

Dalgarno and McCray (31) have discussed the role of fine structure transitions in the thermal balance of the ISM. In the diffuse ISM, excitation of these transitions will occur principally in collisions with electrons and hydrogen atoms.

5.1. Excitation of fine structure transitions by electron impact

The rate (cm^3 s^{-1}) of collisional deexcitation of the upper level j to the lower level i may be written

$$q(j{\rightarrow}i) = 8.63 \times 10^{-6} \frac{T(j,i)}{\omega_j \ T^{\frac{1}{2}}} \tag{5.1}$$

where

$$T(j,i) = \int_{0}^{\infty} \Omega(j,i) \ e^{-x_j} \ dx_j \tag{5.2}$$

and

$$x_j = E_j/(kT) \tag{5.3}$$

In (5.2), $\Omega(j,i)$ is the collision strength for the transition, a dimensionless quantity, symmetric in i and j, and related to the cross-section through

$$\pi\Omega(j,i) = k_j^2 \ \omega_j \ Q(j{\rightarrow}i), \tag{5.4}$$

where

$$k_j^2 = 2E_j \tag{5.5}$$

and ω_j is the statistical weight of level j. Atomic units are employed in Equs. (5.4) and (5.5). It may be seen from Equ. (5.2) that the parameter T and the collision strength, Ω, are identical if the latter is independent of the collision energy, E.

In Table 3, we list atomic data for transitions within the ground terms of $C^+(2s^2 \ 2p \ {}^2P^0)$ and $Si^+(3s^2 \ 3p \ {}^2P^0)$, which are likely to play a role in the thermal balance of the diffuse ISM. Collisional deexcitation can assume importance, relative to spontaneous radiative decay, if the electron density, n_e, is sufficiently high, particularly in the case of C^+. To my knowledge, no recent calculation of $\Omega({}^2P_{1/2}{}^0, \ {}^2P_{3/2}{}^0)$ for Si^+ exists. The estimation of Seaton (32) derives from the application of quantum defect theory, whereas Blaha (33) used the Coulomb-Born approximation, with exchange.

An important addition to the pool of atomic data available for studies of the interstellar medium has come through recent work by Nussbaumer and Storey (37). These authors used the close coupling approximation to calculate collision strengths for transitions within and between the four lowest terms of Fe^+. The results of their calculations of $\Omega({}^6D_{9/2}, \ {}^6D_{7/2})$ and $\Omega \ {}^6D_{9/2}, \ {}^6D_{5/2})$ are much smaller than earlier estimates (38).

Ion	Transition $i - j$	$(E_j - E_i)/k$ (K)	$\Omega(i,j)$	$A(j \to i)$ s^{-1}
C^+	$^2P_{1/2}{}^0 - {}^2P_{3/2}{}^0$	92	$\begin{cases} 1.3\ (32) \\ 1.3\ (33) \\ 1.4\ (34) \end{cases}$	$2.29(-6)\ (35)$
Si^+	$^2P_{1/2}{}^0 - {}^2P_{3/2}{}^0$	413	$\begin{cases} 7.7\ (32) \\ 3.8\ (33) \end{cases}$	$2.17(-4)\ (36)$

Table 3. Collision strengths, Ω, for excitation of fine structure transitions in the $^2P^0$ ground terms of C^+ and Si^+; $A(j \to i)$ is the corresponding spontaneous radiative transition probability.

The most recent calculation of collision strengths for transitions within the ground term, ^3P, of O^0 is by Le Dourneuf and Nesbet (39). For transitions within the ground state multiplets of positive ions (configurations $2p^q$, $q = 1,2,4,5$), reference should be made to Saraph et al. (34). More accurate values for Ne^+ (40) and O^{2+} (41) have since been obtained. The latter is a particularly elaborate calculation, based upon the R-matrix method, and incorporating resonance contributions to the collision strengths. The parameter T, defined by Equ. (5.2), is tabulated by Aggarwal et al. for a range of values of the kinetic temperature, T.

As T increases, transitions to excited terms of the ground configuration and intercombination transitions to excited configurations become increasingly important. Atomic data for these transitions have been reviewed by Flower (42) in the context of the emission line spectra of planetary nebulae.

5.2. Excitation of fine structure transitions by hydrogen

Dalgarno and McCray (31) note the importance of the reactions

$$C^+(^2P_{1/2}{}^0) + H^0 \to C^+(^2P_{3/2}{}^0) + H^0 \qquad (5.6)$$

and

$$O^0(^3P_2) + H^0 \to O^0(^3P_{1,0}) + H^0 \qquad (5.7)$$

to the thermal balance of the ISM. Elaborate calculations of the cross-section for (5.6) have since been made by Bazet et al. (43), who used a semi-classical approximation, and by Launay and Roueff (44), who employed the quantum mechanical close coupling approximation. Launay and Roueff tabulate values of the cooling

T(K)	10^{24} L(T) erg cm^3 s^{-1}		
	(1)	(2)	(3)
10	0.002		0.16
20	0.18		0.68
30	0.88		1.24
40	1.93		1.73
50	3.12		2.14
60	4.29		2.54
70	5.38		2.76
80	6.39		3.00
90	7.31		3.21
100	8.14	0.19	3.41
150	11.3		
200	13.3	1.07	4.70
250	14.7		
300	15.8	2.13	5.58
400	17.5	3.15	6.28
500	18.7	4.11	6.88
600	19.6	5.00	7.41
800	21.2	6.58	8.32
1000	22.5	7.97	9.12

Table 4. Cooling rate coefficients, L, as a function of kinetic temperature, T, for fine structure excitation of (1) C^+, (2) O^0, and (3) C^0 by H^0.

function, $L(T)$, defined as

$$L(T) = \left(\frac{8kT}{\pi M}\right)^{\frac{1}{2}} (E_{1/2} - E_{3/2}) \int_0^\infty x_{1/2} \, Q_{1/2 \to 3/2} \, (x_{1/2}) \, e^{-x_{\frac{1}{2}}} \, dx_{1/2}$$

(5.8)

where M is the reduced mass of the C^+-H^0 system, and

$$x_{\frac{1}{2}} = E_{\frac{1}{2}} / (kT).$$

(5.9)

Their results are reproduced in Table 4.

Launay and Roueff (45) have made analagous calculations of cross-sections for reaction (5.7) and for

$$C^0(^3P_0) + H^0 \to C^0(^3P_{1,2}) + H^0.$$

(5.10)

The corresponding cooling functions,

$$L(T) = \left(\frac{8kT}{\pi M}\right)^{\frac{1}{2}} \sum_{J'} (E_J - E_{J'}) \int_0^\infty x_J \, Q_{J \to J'}(x_J) \, e^{-x_J} \, dx_J,$$

(5.11)

where J denotes the ground state, are also listed in Table 4.

The cooling functions for fine structure excitation of C^+ by ortho- and para- H_2 have been calculated by Flower and Launay (46). The main uncertainties in these calculations are associated with the interaction potentials employed, particularly at short range.

ACKNOWLEDGEMENTS

I am indebted to Drs. H. Nussbaumer, P.J. Storey and P. Valiron for discussions of topics covered in this review and to Professor A. Omont for the hospitality of the Groupe d'Astrophysique de l'Université de Grenoble, where the manuscript was completed.

REFERENCES

(1) Field, G.B., and Steigman, G.: 1971, Astrophys. J. 166, 59.
(2) Chamberlain, J.W.: 1956, Astrophys. J. 124, 390.
(3) Williams, R.E.: 1973, Mon. Not. Roy. Astron. Soc. 164, 111.
(4) Bates, D.R., and Moiseiwitsch, B.L.: 1954, Proc. Phys. Soc.
      67, 805.
(5) Mott, N.F., and Massey, H.S.W.: 1965, The Theory of Atomic
      Collisions, Oxford University Press, London.
(6) Butler, S.E., and Dalgarno, A.: 1980, Astrophys. J. 241, 838.
(7) Bates, D.R.: 1960, Proc. Roy. Soc. A257, 22.
(8) McCarroll, R., and Valiron, P.: 1976, Astron. Astrophys. 53, 83.

(9)   Gargaud, M., Hanssen, J., McCarroll, R., and Valiron, P.:
      1981, J. Phys. B: Atom. Molec. Phys. 14, 2259.
(10) Butler, S.E., Heil, T.G., and Dalgarno, A.: 1980, Astrophys.
      J. 241, 442.
(11) Gargaud, M., McCarroll, R., and Valiron, P.: 1982, Astron.
      Astrophys. 106, 197.
(12) Watson, W.D., and Christensen, R.B.: 1979, Astrophys. J. 231,
      627.
(13) Chambaud, G., Launay, J.M., Levy, B., Millie, P., Roueff, E.,
      and Tran Minh, F.: 1980, J. Phys. B.: Atom. Molec. Phys.
      13, 4205.
(14) Burgess, A.: 1965, Astrophys. J. 141, 1588.
(15) Storey, P.J.: 1981, Mon. Not. Roy. Astron. Soc. 195, 27P.
(16) Aldrovandi, S.M.V., and Pequignot, D.: 1973, Astron.
      Astrophys. 25, 137.
(17) Burgess, A.: 1964, Mem. Roy. Astron. Soc. 69, 1.
(18) Flower, D.R., and Seaton, M.J.: 1969, Computer Phys. Comm.
      1, 31.
(19) Harrington, J.P., Seaton, M.J., Adams, S., and Lutz, J.H.:
      1982, Mon. Not. Roy. Astron. Soc. 199, 517.
(20) Le Dourneuf, M., Vo Ky Lan, and Hibbert, A.: 1976, J. Phys. B:
      Atom. Molec. Phys., 9, L359.
(21) Le Dourneuf, M., Vo Ky Lan, and Zeippen, C.J.: 1979, J. Phys. B:
      Atom. Molec. Phys. 12, 2449.
(22) Burke, P.G., and Taylor, K.T.: 1979, J. Phys. B: Atom. Molec.
      Phys. 12, 2971.
(23) Taylor, K.T., and Burke, P.G.: 1976, J. Phys. B: Atom. Molec.
      Phys. 9, L353.
(24) Luke, T.M.: 1982, J. Phys. B: Atom. Molec. Phys. 15, 1217.
(25) Pradhan, A.K.: 1979, J. Phys. B: Atom. Molec. Phys. 12, 3317.
(26) Pradhan, A.K.: 1980, Mon. Not. Roy. Astron. Soc. 190, 5P.
(27) Drew, J.E., and Storey, P.J.: 1982, J. Phys. B: Atom. Molec.
      Phys., in press.
(28) Tarter, C.B.: 1971, Astrophys. J. 168, 313.
(29) Tarter, C.B.: 1973, Astrophys. J. 181, 607.
(30) Gould, R.J.: 1978, Astrophys. J. 219, 250.
(31) Dalgarno, A., and McCray, R.A.: 1972, Ann. Rev. Astron.
      Astrophys. 10, 375.
(32) Seaton, M.J.: 1958, Rev. Mod. Phys. 30, 979.
(33) Blaha, M.: 1969, Astron. Astrophys. 1, 42.
(34) Saraph, H.E., Seaton, M.J., and Shemming, J.: 1969, Phil.
      Trans. Roy. Soc. A 264, 77.
(35) Nussbaumer, H., and Storey, P.J.: 1981, Astron. Astrophys.
      96, 91.
(36) Nussbaumer, H.: 1977, Astron. Astrophys. 58, 291.
(37) Nussbaumer, H., Storey, P.J.: 1980, Astron. Astrophys. 89, 308.
(38) Seaton, M.J.: 1955, Ann. Astrophys. 18, 188.
(39) Le Dourneuf, M., and Nesbet, R.K.: 1976, J. Phys. B: Atom. Molec.
      Phys. 9, L241.
(40) Seaton, M.J.: 1975, Mon. Not. Roy. Astron. Soc. 170, 475.
(41) Aggarwal, K.M., Baluja, K.L., and Tully, J.A.: 1982, Mon. Not.
      Roy. Astron. Soc., in press.

(42) Flower, D.R.: 1983, to be published.
(43) Bazet, J.F., Harel, C., McCarroll, R., and Riera, A.: 1975,
     Astron. Astrophys. 43, 229.
(44) Launay, J.M., and Roueff, E.: 1977, J. Phys. B: Atom. Molec.
     Phys. 10, 879.
(45) Launay, J.M., and Roueff, E.: 1977, Astron. Astrophys. 56, 289.
(46) Flower, D.R., and Launay, J.M.: 1977, J. Phys. B: Atom. Molec.
     Phys. 10, 3673.

INDEX